NATURAL LANDSCAPE AMENITIES AND SUBURBAN GROWTH

METROPOLITAN CHICAGO, 1970-1980

by

Christopher Mueller-Wille

Michelin Travel Publications

UNIVERSITY OF CHICAGO

GEOGRAPHY RESEARCH PAPER NO. 230

1990

Copyright 1990 by

The Committee on Geographical Studies
The University of Chicago
Chicago, Illinois

Library of Congress Cataloging-in-Publication Data

Mueller-Wille, Christopher.
    Natural landscape amenities and suburban growth : metropolitan Chicago, 1970-1980 / by Christopher Mueller-Wille.
       p.   cm. — (Geography research paper ; no. 230)
    Includes bibliographical references (p.   ).
    ISBN 0-89065-136-1
    1. Suburbs—Illinois—Chicago Metropolitan Area. 2. Chicago Metropolitan Area (Ill.)—Population density. 3. Landscape—Social aspects—Illinois—Chicago Metropolitan Area. 4. Land use, Urban—Illinois—Chicago Metropolitan Area. I. Title. II. Series: Geography research paper (Chicago, Ill.) ; no. 230.
HT352.U62C46   1990
307.74'09773'11—dc20   90-10818
                                                                          CIP

Geography Research Papers are available from:

The University of Chicago
Committee on Geographical Studies
5828 South University Avenue
Chicago, Illinois 60637-1583

Meinen Eltern: Vorbild und Leitbild

"...ordnend zu beobachten und verbindend zu denken."
—W. Müller-Wille, *Westfalen*

CONTENTS

List of Figures	vii
List of Tables	xi
Acknowledgments	xi
1. INTRODUCTION	1
2. PERSPECTIVES ON THE STUDY OF SUBURBS AND NATURAL LANDSCAPE FEATURES	7
Suburban Prospectus	7
On Defining Suburbs	11
On Measuring Suburbanization	12
On Excluding Environmental Factors	14
Choices of Ideology and Methodology	19
3. CONCEPTS OF THE NATURAL ENVIRONMENT AS AMENITY IN AN URBAN SOCIETY	25
Cultural Values and Predispositions	27
City System Evolution and Metropolitan Development	32
4. METHODOLOGY, DATA BASE, AND OPERATIONAL DEFINITIONS	37
Method of Analysis	37
Origin and Use of Density Functions	38
Estimation Form of the Modified Density Function	41
Selection of Variables and Data Sources	43
Spatial Matrix of the Study Area	45
The Variables and Their Operational Definition	47
5. NATURAL LANDSCAPE AMENITIES AND POPULATION DENSITIES: SECTORAL PROFILES	61
The North Sector	64
The North-Shore Margin	69
The Northwest Sector	74
The West Sector	79
The Southwest Sector	84
The South Sector	89
Sectoral Summary and Overview	94

6. REGIONAL POPULATION DENSITIES AND PATTERNS OF CHANGE   99
    Regional Gradients   99
    Suburban Densities and Patterns of Change   102
7. SUMMARY AND OUTLOOK   117
Appendix A: Pattern of Spatial Aggregation within a Township   122
Appendix B: Descriptive Statistics   123
Bibliography   133
Index   151

FIGURES

1. Landforms of the Chicago metropolitan area. 4
2. The geometry of the rectangular land survey in northeastern Illinois. 46
3. The distribution of suburban forests. 48
4. The distribution of suburban nature parks and preservation areas. 50
5. The distribution of suburban golf courses. 51
6. The distribution of suburban lakes and reservoirs. 52
7. The distribution of suburban rivers and tributaries. 53
8. Relative variation in elevation within designated four-section areas. 54
9. Maximum elevation within designated four-section areas. 55
10. Minimum elevation within designated four-section areas. 56
11. Centroids of designated four-section areas within three miles of suburban commuter railroad stations. 58
12. Centroids of designated four-section areas within two miles of limited-access expressways and interstates. 59
13. Relationship between suburban population density and linear distance in the north sector, 1970 and 1980. 67
14. Density gradient surfaces of the north-shore margin, 1970 and 1980. 72
15. Relationship between suburban population density and linear distance along the north-shore margin, 1970 and 1980. 73
16. Relationship between suburban population density and linear distance in the northwest sector, 1970 and 1980. 76
17. Relationship between suburban population density and linear distance in the west sector, 1970 and 1980. 83
18. Relationship between suburban population density and linear distance in the southwest sector, 1970 and 1980. 86
19. Relationship between suburban population density and linear distance in the south sector, 1970 and 1980. 92
20. The distribution of suburban population in 1970. 103
21. The distribution of suburban population in 1980. 104
22. The regional pattern of relative increase in suburban population density between 1970 and 1980. 107

FIGURES

23. The regional pattern of relative decrease in suburban population density between 1970 and 1980.   108
24. The regional pattern of absolute losses in suburban population counts between 1970 and 1980.   109
25. The regional pattern of absolute gains in suburban population counts between 1970 and 1980.   110
26. Generalized, cross-sectional model of suburban population change in the Chicago metropolitan area, 1970-1980.   111
27. Changing levels of suburban population density in the north sector for 1970 and 1980.   112
28a. Changing levels of suburban population density in the north-shore margin for 1970 and 1980.   113
28b. Changing levels of suburban population density with increasing distance from Lake Michigan within the north-shore margin for 1970 and 1980.   113
29. Changing levels of suburban population in the northwest sector for 1970 and 1980.   114
30. Changing levels of suburban population density in the west sector for 1970 and 1980.   114
31. Changing levels of suburban population density in the southwest sector for 1970 and 1980.   115
32. Changing levels of suburban population density in the south sector for 1970 and 1980.   115

TABLES

1. Sectoral dimensions and population statistics. 63
2. Sectoral mean and extreme values of six natural landscape variables. 64
3. Determinants of population density in the north sector, 1970. 65
4. Determinants of population density in the north sector, 1980. 66
5. Determinants of population density along the north-shore margin, 1970: coefficients and (standard errors) of landscape and access variables. 70
6. Determinants of population density along the north-shore margin, 1980. 71
7. Determinants of population density in the northwest sector, 1970. 74
8. Determinants of population density in the northwest sector, 1980. 75
9. Determinants of population density in the west sector, 1970. 80
10. Determinants of population density in the west sector, 1980. 81
11. Determinants of population density in the southwest sector, 1970. 85
12. Determinants of population density in the southwest sector, 1980. 87
13. Determinants of population density in the south sector, 1970. 90
14. Determinants of population density in the south sector, 1980. 91
15. Significant determinants of sectoral population densities and relative shifts in accounted variation from 1970 to 1980. 95
16. Changes in the coefficients of determination. 96
17. Determinants of population density in Chicago's suburban region, 1970. 100
18. Determinants of population density in Chicago's suburban region, 1980. 101

ACKNOWLEDGMENTS

The credit for whatever is commendable in this book I share with those named [below] and with many others. The blame for defects will be wholly my own.

—D. Greenhood, *Mapping*

    This study would not have been finished if it had not been for the assistance of several people. Charles Metalitz of the Northeastern Illinois Planning Commission (NIPC) provided the data in accessible and compatible formats. Professor Brian Blouet at Texas A&M University secured the financial support to purchase NIPC's 1980 updates. Judith Duncan was quick in locating and shipping much-needed topographic maps. My special gratitude extends to Professors Michael Conzen and Marvin Mikesell at the University of Chicago. Both have shown the flexibility, patience, and understanding needed to see this study through to its end. Finally, Carol Saller as the editor of the Geography Research Papers series was instrumental in readying the manuscript for publication.

    This is also the right place to acknowledge the debt owed to those who gave support beyond the practical advice toward research and writing. My particular thanks go to: Professor Daniel Arreola, who took interest even when faced with only fragmentary evidence; Professor Brian Berry, who in his inimitable fashion proffered a spirited, no-nonsense approach toward overcoming procrastination and timidity; Professor Karl Butzer, who reassured with friendly advice and unwavering trust; Professor Chauncy Harris, who gave encouragement. My special gratitude is reserved for Harriet S. Platt, for without her help and compassion, without her friendship, my academic odyssey would never have taken me to Chicago.

    My greatest debt is owed to Catherine Mueller-Wille. Because she had faced a similar situation at the University of Chicago, she understood best. Because she had successfully fought through it herself, she set the standard and gave hope. Her love and support have made the difference.

Chapter 1

INTRODUCTION

The suburb is America's greatest success, in settlement terms.
—J.E. Vance, *This Scene of Man*

In this study I examine the spatial relationship between selected natural landscape features and suburban population densities in the Chicago metropolitan area, and interpret the changes that occurred between 1970 and 1980. My basic assumption is that natural landscape features, as amenities, play discernible roles in shaping the physical form of contemporary suburban development, its direction, intensity, and reach. Environmental conditions are often studied as disamenities, negative factors akin to "linear distance," which is the catch-all term for a variety of social and mostly economic constraints. Instead, I focus on those natural landscape elements that may counteract the negative effect of traditionally invoked variables on residential densities within metropolitan areas.

The physiographic features and conditions considered in this study are the visible, surficial, and fairly permanent components of the natural landscape such as relative relief, vegetation cover, and surface water. Each encompasses a host of landscape aspects that are valued and desirable locational attributes within the suburban housing market. I propose that some of these natural landscape elements, in addition to historically persistent commuter rail corridors and current freeway access, account at least in part for the spatial variation in Chicago's suburban population densities and underlie both the redistribution of suburban populations and the patterns of residential expansion along the metropolitan periphery.

As suburban regions go, the suburban portion of the six-county Chicago metropolitan area appears to hold little promise as a subject for a study that intends to analyze the spatial variability of natural landscape features and their effects on population densities. Without much effort one could name a number of metropolitan areas endowed with greater physiographic diversity that is more visually striking and more easily identified than Chicago's relatively modest suburban environs. However despite their relative subtlety, one can observe a variety of environmental settings and features that provide sufficient visual contrast and local identity within the region.

The region is characterized by an ensemble of landforms that is typical of the glacially overformed landscapes in this part of the country.[1] Sweeping arcs of terminal and recessional Pleistocene moraines and elongated ridges of sandy, post-glacial beach deposits alternate with near-level lacustrine silt or clay beds as well as glacial outwash plains, fans, and sags (figure 1). The moraines have an elevated, hummocky topography with local pockets of sand, gravel, and peat in addition to glacial till. Even though the presettlement vegetational division of prairie and forest is still reflected in the region's two major soil associations, the indigenous vegetational regimes have largely been supplanted during the last one hundred and fifty years by agricultural and urban vegetation.[2]

Chicago's modern suburban region stretches across two broadly defined physiographic landscapes. The northern and northwestern portion of the region shows greater local relief, is dotted by numerous lakes and is still covered by extensive forests. In contrast, the area immediately surrounding Chicago and to the west and south of the city has a nearly level or gently rolling upland topography dissected by wooded stream valleys that generally follow a south-southwestern course. Any physiographic description of the Chicago region would remain incomplete if it were not to mention Lake Michigan. It is the region's most attractive and best-known environmental asset and therefore will warrant special consideration in the analysis.

To some researchers Chicago represents the archetypal American city; others view it as a cosmopolitan city.[3] From the outset it has played an in-

---

[1] *Quaternary Deposits of Illinois*, compiled by J.A. Lineback (Springfield: Illinois State Geological Survey, 1979).

[2] *General Soil Map of Illinois* (Urbana-Champaign: University of Illinois Agricultural Experiment Station, 1982). Relics of native prairie and forest ecosystems can be found in several of the region's forest preserves and conservation areas. See J.A. Schmid, *Urban Vegetation*, University of Chicago Department of Geography Research Paper no. 161 (Chicago: University of Chicago Department of Geography, 1975), pp. 12-14.

[3] H.M. Mayer, "Chicago als Weltstadt," in *Zum Problem der Weltstadt*, ed. J. H. Schultze (Berlin: De Gruyter, 1959), pp. 83-111. Chicago's seductive prominence led a group of geographers to proclaim that the city is "less a place than it is a process." B.J.L. Berry et al., *Chicago: Transformation of an Urban System* (Cambridge, Mass.: Ballinger, 1976), p. 1. Aca-

fluential part in American urban history. It has been the source and object of numerous conceptual and theoretical models of urban form, growth, and development.[4] As a result, a great deal of information is available on Chicago's historical growth and present condition. During the last two decades the Northeastern Illinois Planning Commission (NIPC) has begun to compile a series of data files detailing population counts, land uses, and environmental inventories. For the most part, these files cover only the suburban portion of the Chicago metropolitan area. What makes the files unique is that they share a common geographic referencing system with a spatial resolution that not only surpasses most other existing sources of information (e.g., U.S. Bureau of the Census counts and estimates), but also is internally consistent and has not changed through time.

The importance of natural landscape amenities has been discussed with increasing frequency, particularly in studies dealing with interregional migration, regional growth in general, and the recent growth of nonmetropolitan areas. If natural environmental attributes are important variables in understanding long-distance migration, why should they not play a role in short-distance moves? In many studies natural amenities are given perfunctory treatment owing in part to the relative paucity of detailed environmental data. Whereas surrogate data can be used in regional comparisons, the general scarcity or lack of appropriate information with sufficient local resolution has discouraged studies of individual metropolitan areas. For this reason, the NIPC data files provided both incentive and opportunity to conduct a metropolitan-sized study that examined the spatial association of natural landscape features and suburban population densities.

This study does not attempt to redefine the division between Chicago and its suburban communities, nor does it attempt to identify the more subtle, localized differences in urban morphology for the entire metropolitan area. The distinction between Chicago and its suburbs follows the conventional standard of accepting municipal boundaries of the central city as

---

demic hyperbole such as the above is understandable, for the city's apparent magnetic spell has been pervasive. During its rapid rise to cosmopolitan metropolis Chicago not only attracted immigrant labor and capital investments, it also engaged and challenged the creativity of poets and writers (Carl Sandburg, Bertolt Brecht), architects (Frank Lloyd Wright, Ludwig Mies van der Rohe), and urban planners (Frederick Law Olmsted, George Pullman, Daniel H. Burnham, William S. Whyte).

[4] One of the more readily recognized groups of models originated with the "human ecology school of Chicago" as discussed by J.A. Agnew, J. Mercer, and D.E. Sopher's introduction to *The City in Cultural Context*, ed. J.A. Agnew, J. Mercer, and D.E. Sopher (Boston: Allen and Unwin, 1985), p. 12. Similar classifications are expressed by J.W.R. Whitehand, "Urban Morphology," in *Historical Geography: Progress and Prospect*, ed. M. Pacione (London: Croom Helm, 1987), pp. 250-256.

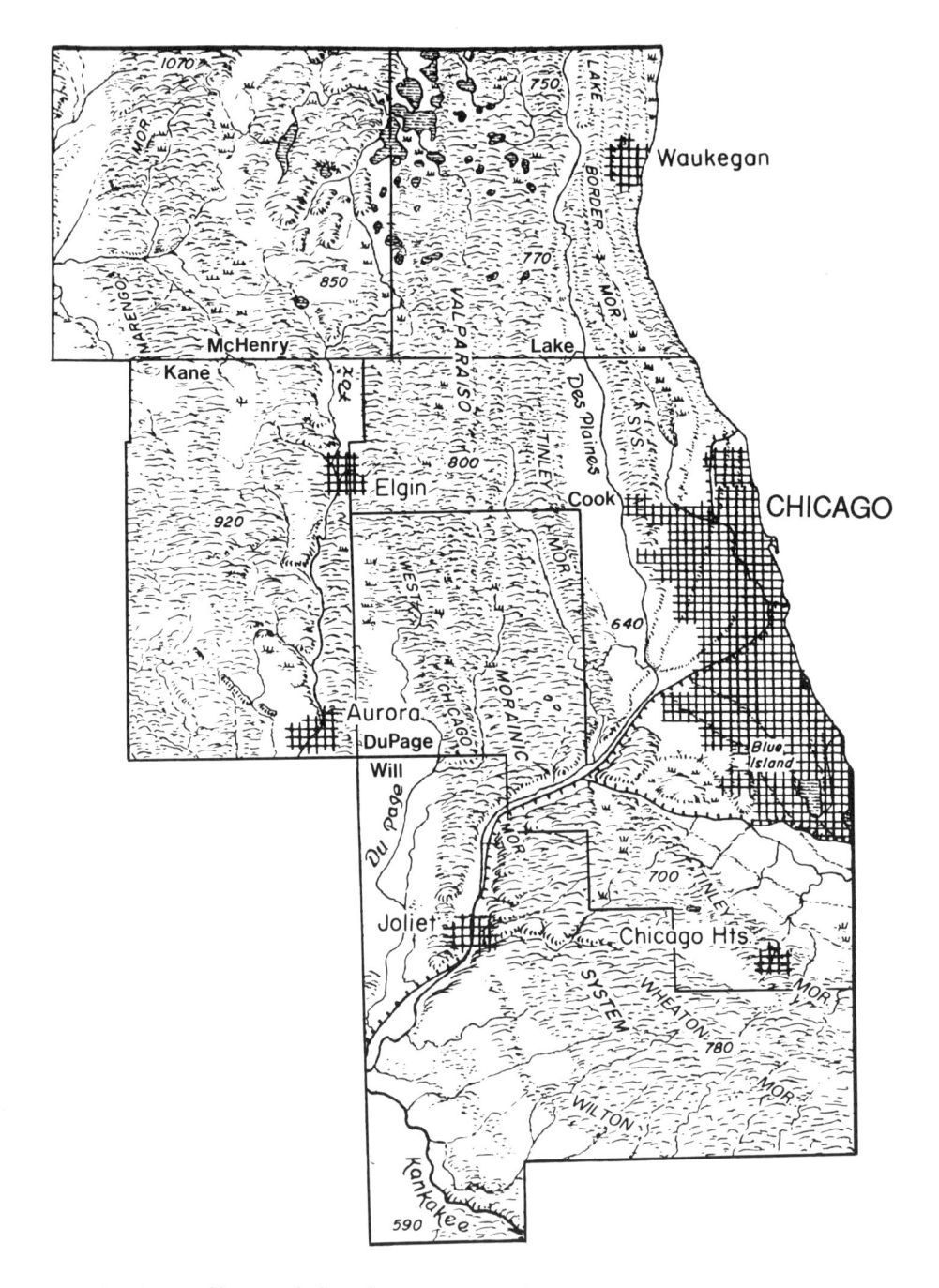

Fig. 1. *Landforms of the Chicago metropolitan area. After* Landforms of Illinois, 1980.

the dividing line. Thus the study area lies outside the city of Chicago proper and the two imbedded municipalities of Harwood Heights and Norridge. The analysis of suburban population densities includes the remaining portion of Cook County and the five surrounding counties of Lake, McHenry, Kane, DuPage, and Will (figure 1).

The study begins, in the following chapter, with a broad perspective on suburban development in the United States. The notion of suburban living in the United States draws from the opposites of urban culture and pastoral nature. A review of the literature suggests that the study of American suburbanization has been approached from various directions and that population size and density together with natural landscape features have not been considered basic elements in this process. Natural landscape features, especially, have been either disregarded or only obliquely incorporated recent urban geographic research.

The third chapter discusses research efforts that explicitly recognize the significance of natural environmental factors and conditions and seek to weigh their influence on the distribution of urban populations. The chapter's chief objective is to delineate a series of conceptual frameworks that define the parameters of this study. The chapter's focus rests squarely on the geographical and historical contexts within which environmental features emerge as crucial locational amenities that have transformed the national system of cities and fostered population growth at the metropolitan level.

The fourth chapter introduces the principal method of analysis. In it, I discuss the criteria of selecting relevant landscape features, present the study's data base, and discuss the operational definition of each variable. A series of maps portray the regional distribution of each independent variable, except for linear distance, and illustrate the subdivision of the Chicago suburban region into five sectors.

The fifth chapter analyzes the sectoral patterns of population density in terms of natural endowment and metropolitan accessibility. Three questions are central to the analysis:

1. To what extent have environmental factors counteracted the constraints of distance and access?
2. Have the specific effects of individual natural landscape features been of comparable strength across the region, both qualitatively and quantitatively?
3. What have been the most noticeable changes in suburban population distribution between 1970 and 1980 and how much change can be attributed to natural landscape features and conditions?

The sixth chapter presents a regional overview and closes with a cartographic exploration of population change across the entire suburban region. Maps of population gains and losses offer a visual comparison and counterpoint to the preceding statistical analyses. Not mere cross-sections frozen in time, they illustrate the magnitude and geographic distribution of change in both relative and absolute terms. The final chapter contains a summary of the study's principal findings and discusses the scope of related future research.

Chapter 2

PERSPECTIVES ON THE STUDY OF SUBURBS AND NATURAL LANDSCAPE FEATURES

In the United States, it is almost a truism to observe that the dominant residential pattern is suburban.

—K.T. Jackson, *Crabgrass Frontier*

There is an environment of the mind as real as sticks and stones which, given scope, can remake the environment of the land. Mindscape America is not a figment: it is the image of America worked out in the American scene. This is not a contrived thing, but real, but powerful.

—J.W. Watson, *Social Geography of the United States*

Suburban Prospectus

To many Americans the move to the suburbs represents search: they seek to realize a residential ideal that emphasizes living in socially tranquil communities and environmentally pleasant surroundings.[1] To some it rep-

---

[1] The general themes expressed here in condensed form are part and parcel of the majority of descriptions and analyses concerned with the origin and growth of the American suburb. A comprehensive overview of suburban history and ideology is provided by K.T. Jackson, *Crabgrass Frontier: Suburbanization in the United States* (New York: Oxford University Press, 1985). Also see M.S. Marsh and S. Kaplan, "The Lure of the Suburbs," in *Suburbia: The American Dilemma*, ed. P.C. Dolce (Garden City: Anchor Books, 1976), pp. 37-58; D. Nicholson-Lord, *The Greening of the Cities* (London: Routledge and Kegan Paul, 1987); C. Perin, *Everything in Its Place: Social Order and Land Use in America* (Princeton: Princeton University Press, 1977); J.A. Tarr, "From City to Suburb," in *American Urban History*, ed. A.B. Callow, Jr. (New York: Oxford University Press, 1973), pp. 202-212; Y.F. Tuan, *Topophilia* (Englewood Cliffs: Prentice-Hall, 1974); J.W. Watson, "The Image of Nature in America," in *The American Environment*, ed. J.W. Watson and T. O'Riordan (New York: Wiley, 1976); J.W. Watson, *Social Geography of the United States* (London: Longman, 1979). The mind-boggling diversity of the academic literature on suburbia is plainly documented in the annotated

resents flight: they escape from a central city that they perceive to be deteriorating both socially and environmentally. Suburbs serve the dual purpose of shutting out the unwanted and cocooning its denizens.[2] This Janus-faced concept of suburbia involves "both planning type and state of mind based on imagery and symbolism."[3] Suburbia as haven promises to protect its citizens and welcomes those well "on their way to Americanization."[4] Suburbia as promise synthesizes the mythical opposites of a rural Eden and an urban inferno. Suburbia as synthesis creates a "middle landscape" between the industrial city of immigrants and the pioneers' pristine wilderness;[5] it

---

bibliography by J. Zikmund II and D.E. Dennis, *Suburbia: A Guide to Information Sources*, Urban Studies Information Guide Series vol. 9 (Detroit: Gales Research Co., 1979).

[2] C.A. O'Connor, *A Sort of Utopia: Scarsdale, 1881-1981* (Albany: State University of New York Press, 1983).

[3] Jackson, *Crabgrass Frontier*, pp. 4-5. The special role of new towns and greenbelt towns is discussed in I.L. Allen, ed., *New Towns and the Suburban Dream* (Port Washington, N.Y.: Kennikat, 1977). Also see R.B. Zehner, *Indicators of the Quality of Life in New Communities* (Cambridge, Mass.: Ballinger, 1977); and D.C. Klein, ed., *Psychology of the Planned Community: The New Town Experience* (New York: Human Sciences Press, 1978).

[4] E.W. Burgess, "The Determination of Gradients in the Growth of the City," *Publications of the American Sociological Society* 21 (1927): 178.

[5] Tuan, *Topophilia*, p. 108, credits the term "middle landscape" to L. Marx, *The Machine in the Garden: Technology and the Pastoral Ideal in America* (New York: Oxford University Press, 1964), p. 5. The term was subsequently used by W. Zelinsky, *The Cultural Geography of the United States* (Englewood Cliffs: Prentice-Hall, Foundations in Geography Series, 1973), p. 64; and by Marsh and Kaplan, "The Lure of the Suburbs," p. 54. The zone of transition from city to farmland had been called "borderland" by a number of nineteenth-century writers. The term is resurrected by J.R. Stilgoe, *Borderland: Origins of the American Suburb, 1820-1939* (New Haven: Yale University Press, 1988). A decidedly less romantic view is embodied in the term "outer city," which denotes the coming of age of the present-day American suburb as interpreted by P.O. Muller, *The Outer City: The Geographical Consequences of the Urbanization of the Suburbs*, Association of American Geographers Resource Paper no. 75-2 (Washington, D.C.: Association of American Geographers, 1976). The concept of the rural-urban fringe has been used to encompass the entire spectrum of settlements outside the central city from satellite towns and suburbs to exurbs; see R.E. Pahl, *Urbs in Rure: The Metropolitan Fringe in Herfordshire*, London School of Economics and Political Sciences Geographical Paper no. 2 (London: London School of Economics and Political Sciences, 1964); R.J. Pryor, "Defining the Rural Urban Fringe," in *Internal Structure of the City*, ed. L.S. Bourne (New York: Oxford University Press, 1972), pp. 59-68; R.A. Kurtz and J.B. Eicher, "Fringe and Suburb: A Confusion of Concepts," *Social Forces* 37 (1958): 32-37. Today largely abandoned in the United States, it is still used by Canadian urban geographers, although in practice it has been given the more restricted definition as a zone of transition that is relatively unstable in terms of both spatial and temporal duration; see K.B. Beesley and L.H. Russwurm, eds., *The Rural-Urban Fringe: Canadian Perspectives*, York University Department of Geography, Geographical Monograph no. 10 (Toronto: York University Department of Geography, 1981). The fringe-belt concept is solely associated with the works of M.R.G. Conzen in England and provides the link to similar studies in Germany; see M.R.G. Conzen, *The Urban Landscape: Historical Development and Management: Papers by M.R.G. Conzen*, ed. J.W.R. Whitehand, Institute of British Geographers Special Publication no. 13 (London: Academic Press, 1981);

blends the amenities of a bucolic life with the essentials of an urban existence. Suburbia as urban landscape is "the city in the garden" where winding lanes, cul-de-sacs, and detached single-family homes can provide both domestic seclusion and pastoral openness.[6]

The quintessential suburban settlement in the United States features spread-out, horizontal, and scattered residential development. For this reason, some of the less extreme features of the natural environment can become important visual and integral elements of the suburban landscape simply because they are generally scarce and overwhelmed within the older and crowded central city. The relatively subtle, localized natural features arguably are more noticeable in suburban areas and may help differentiate nearly interchangeable suburban communities and neighborhoods. In contrast, when considering such dominant regional features as the Rocky Mountains in the Denver area or Puget Sound in the Seattle-Tacoma area, environmentally based distinctions between central-urban and suburban locations may be altogether irrelevant.

The inferred attraction of natural landscape features is most directly mirrored by the names of suburban communities and residential subdivisions. If desirable landscape features are not entirely accommodated, emphasized, or preserved by deliberate physical design, a suburb's name will conjure up at least one or two natural landscape elements. If not natural features, then a suburb's name will suggest the idyllic serenity of quaint agrarian estates and villages.[7]

W. Hartke, "Die Sozialbrache als Phänomen der geographischen Differenzierung der Landschaft," *Erdkunde* 10 (1956): 257-269; K. Ruppert, "Zur Definition des Begriffes 'Sozialbrache,'" *Erdkunde* 11 (1957): 226-231. Despite these variations in terminology and definition there is a clear understanding that settlement surrounding a central city represents a visibly different (i.e., mappable) landscape type.

[6] J.E. Vance, "California and the Search for the Ideal," *Annals of the Association of American Geographers* 62 (1972): 185. Also see D. Schaffer, *Garden Cities for America: The Radburn Experience* (Philadelphia: Temple University Press, 1982; F. Ermuth, *Residential Satisfaction and Urban Environmental Preferences*, York University Department of Geography, Geographical Monograph no. 3 (Toronto: York University Department of Geography, 1974); R. Kaplan, "The Green Experience," in *Humanscape: Environments for People*, ed. S. Kaplan and R. Kaplan (Ann Arbor: Ulrich's Books, 1982), pp. 186-193. The instinctive need to observe without being seen has been described in greater detail by J. Appleton, *The Experience of Landscape* (New York: Wiley, 1975).

[7] In 1978 there were 260 separate municipalities in the Standard Metropolitan Statistical Area (SMSA) outside Chicago and its two imbedded independent communities. Only 43 (16.5%) were cities in both name and legal standing. A total of 154 municipalities (59.2%) made direct reference to physiographic features and rural or small-town status. Vegetational features (29.6%) were most often used as part of a community's name. Topographic references followed with 22.3% and hydrographic features with 16.5%. Combinations of two categories were common; however names such as Fox River Valley Gardens remain the exception. On the

Such general notions of suburbia are variations of a basic and enduring theme of American urban history, a theme that makes natural environmental features, imagined or real, a touchstone of suburban development. It follows that residential density and natural environment are significant, if partial, structural elements that support a compound, fluid concept of suburbia. To what extent they are governed by or in turn influence societal structure, its institutions, and its agents is by no means clearly understood. Yet, one can without great difficulty acknowledge that each occupies a distinct position within the broader cultural context of contemporary American society. Population density and natural environment, an otherwise antithetical pair, have been fused into a simple scale that to some extent measures a community's success or failure of projecting and sustaining an ultimately suburban ambience. How they combine to direct and shape the present patterns of suburban population distribution is the subject matter of this study.

As simple truism this view smacks of facile, even faddish stereotype that oversimplifies suburban diversity and romanticizes unrealistic suburban ideals. As such it may have dissuaded urban geographers from examining the relationship between natural landscape and the spread of suburban settlement within a metropolitan region. It is, however, equally true that the linkage persists. The question whether as illusion or fact is irrelevant, since this notion is as actively promoted and exploited by developers, planners, lenders, and politicians as it is willingly embraced and defended by suburban residents and civic groups.[8] To disqualify this notion merely on grounds of it being vague and superficial is shortsighted. Geographers have a long-standing commitment to look at both natural and cultural elements that together create distinctive cultural landscape types. An area's physiographic endowment and population distribution are the basic building blocks with which we can begin solving the puzzle of cultural landscape for-

---

political fragmentation of Chicago's suburbs see M.P. Conzen, "American Cities in Profound Transition: The New City Geography of the 1980s," *Journal of Geography* 82 (1983): 100. On the relationship between a community's name and its social status see P. deVise, "The Geography of Wealth and Poverty in Suburban America: 1979 to 1985," City Club of Chicago, Chicago Regional Inventory Working Paper II.90, March 1987, photocopy. The crass discrepancy between advertised image and reality as well as the relationship between house prices and preservation of natural settings are examined by B.A. Weightman, "Arcadia in Suburbia: Orange County, California," *Journal of Cultural Geography* 2 (1981): 55-69.

[8] J.K. Mitchell, "Adjustment to New Physical Environments beyond the Metropolitan Fringe," *Geographical Review* 66 (1976): 18-31; G.F. Hepner, "An Analysis of Residential Developer Location Factors in a Fast Growing Urban Region," *Urban Geography* 4 (1983): 355-363; M. Bunce, "Rural Sentiment and the Ambiguity of the Urban Fringe," in *The Rural-Urban Fringe*, ed. Beesley and Russwurm, pp. 109-120; H.C. Perkins, "Bulldozers in the Southern Part of Heaven, Parts I and II," *Environment and Planning A* 20 (1988): 285-308 and 435-456.

mation. To analyze side-by-side natural landscape elements and patterns of suburban population density represents the first step in that direction.

On Defining Suburbs

The distribution of population has always been near the center of urban geographic studies. Population size and density provide readily grasped and measured baselines; together or separately, they serve to distinguish rural areas from urban(ized) areas, nonmetropolitan regions from metropolitan regions, suburbs from central cities.[9] Nonetheless, the specific conceptual definition of "suburban" has varied considerably. Always bound to an author's particular theoretical perspective and choice of problem, it has reflected the broader concerns of place and time. Setting aside the need to place the study of urban form squarely into a cultural, historical, and physical context,[10] it is difficult to reject outright the idea that American suburbs constitute a distinct type of settled landscape. In order to delineate any structurally and functionally distinct urban landscapes, geographers have looked at the three major physical categories of ground plan, land-use pattern, and building fabric.[11] For the most part, however, researchers of urban

---

[9] U.S. Bureau of the Census, *1980 User's Guide, Part B* (Washington, D.C.: Government Printing Office, 1982). The bureau does not not use "suburb" as a standard technical term. A counterpoint to rigid terminology and definitions is provided by J.K. Hadden, "Use of *ad hoc* Definitions," in *Sociological Methodology*, ed. E.F. Borgatta (San Francisco: Jossey-Bass, 1968), pp. 276-285.

[10] H. Carter, *An Introduction to Urban Historical Geography* (London: Arnold, 1983), pp. 130-149; J.A. Agnew, J. Mercer, and D.E. Sopher, introduction to *The City in Cultural Context*, ed. J.A. Agnew, J. Mercer, and D.E. Sopher (Boston: Allen and Unwin, 1985), ch. 1; R.A. Mohl, "New Perspectives on American Urban History," in *The Making of Urban America*, ed. R.A. Mohl (Wilmington, Del.: Scholarly Resources, 1988), pp. 293-316.

[11] M.P. Conzen, "Analytical Approaches to the Urban Landscape," in *Dimensions of Human Geography*, ed. K.W. Butzer, University of Chicago Department of Geography Research Paper no. 186 (Chicago: University of Chicago Department of Geography, 1978), pp. 134-158; H. Carter, *The Study of Urban Geography*, 2d ed. (London: Arnold, 1975), pp. 182-183; J.W.R. Whitehand, "Urban Morphology," in *Historical Geography: Progress and Prospect*, ed. M. Pacione (London: Croom Helm, 1987), ch. 9. The morphological and morphogenetic approaches to urban landscapes have a long tradition in central European geography; see J.G. Kohl, *Der Verkehr und die Ansiedelungen der Menschen in ihrer Abhängigkeit von der Gestaltung der Erdoberfläche* (Leipzig: Arnoldsche Buchhandlung, 1841). A recent commentary is provided by G. Pfeiffer, ". . . 'und man sollte J.G. Kohl nicht vergessen!'" in *Mensch und Erde*, Institut für Geographie und Länderkunde and Geographische Kommission für Westfalen Westfälische Geographische Studien no. 33 (Münster: Institut für Geographie und Länderkunde and Geographische Kommission für Westfalen, 1976), pp. 221-236. A survey of more than 400 English, French, and German academic journals showed that only 12 percent of geographical papers on the internal structure of cities were concerned with these three themes of urban morphology; see J.W.R. Whitehand, "Taking Stock of Urban Geography," *Area* 18 (1986): 147-151.

America have followed a less meticulous, more practical approach. Simply put, suburbs are defined as separate, self-governing communities that lie at the outskirts of a larger, central city and are within commuting range.

It may be convenience rather than choice that forged this consensus. Studies that offer comparative analyses and interpretations of suburbia at the national or regional level would certainly be much smaller in number if it were not for the readily available enumeration data collected within and for politically defined entities. Even though the above plain definition is lacking in detail, it does manage to accommodate the American penchant for distrusting authority and cherishing privacy. From this perspective, the political fragmentation of suburban areas—in part brought about by the voting residents' rejection of central city annexation—is the result of choice, in that the deep-seated affection for localism and individualism demands a small, responsive government.[12] Whatever the reasoning, much of what we claim to know today about suburbia in the United States hinges on a politically defined city-suburb dichotomy and not on analyses of detailed field surveys.[13]

On Measuring Suburbanization

During the last fifty years or so suburbs have become the dominant type of settlement in the United States.[14] The country's far-flung, sprawling form of suburban settlement is altogether unique, as the evidence of similar

---

[12] M.A. Goldberg and J. Mercer, *The Myth of the North American City* (Vancouver: University of British Columbia Press, 1986), pp. 139-147; Zelinsky, *The Cultural Geography*, pp. 36-64; R.C. Wood, "The American Suburb," in *Man and the Modern City*, ed. E. Green et al. (Pittsburgh: University of Pittsburgh Press, 1963), pp. 112-121.

[13] The objectives that underlie such painstaking research of urban morphology are outlined by Conzen, "Analytical Approaches." There are of course the detailed sociological case studies of individual suburban communities that have greatly enriched our understanding of the suburban way of life; see B.W. Berger, *Working-Class Suburb: A Study of Auto Workers in Suburbia* (Berkeley: University of California Press, 1971); and H.J. Gans, *The Levittowners: Ways of Life and Politics in a New Suburban Community* (New York: Vintage Books, 1967). The general and persistent lack of in-depth suburban community studies as well as the apparent ambivalence concerning a commonly accepted definition of suburbia are documented by Zikmund II and Dennis, *Suburbia*.

[14] H. Douglass, *The Suburban Trend* (1925; reprint, New York: Arno Press, 1970); R.D. McKenzie, "The Rise of Metropolitan Communities," in *On Human Ecology: Selected Writings*, ed. A.H. Hawley (1933; reprint, Chicago: University of Chicago Press, 1968), ch. 14; C.D. Harris, "Suburbs," *American Journal of Sociology* 49 (1943): 1-13; B.J.L. Berry, *The Human Consequences of Urbanization* (New York: St. Martin's, 1973); P.F. Lewis, "The Galactic Metropolis," in *Beyond the Urban Fringe*, ed. R.H. Platt and G. Macinko (Minneapolis: University of Minnesota Press, 1983), pp. 23-49.

developments abroad remains scant and inconclusive.[15] Despite greatly divergent interpretations of the principal causes and consequences of suburbanization in the United States, there is little disagreement that population size and density are two of the more conspicuous dimensions that separate suburb from central city.

Decentralization and deconcentration are frequently used measures that describe the extent of suburbanization within a metropolitan area,[16] each depicting a different pattern of population redistribution. The first follows municipal boundaries and shows an increasing proportion of the metropolitan population residing outside the central city. Decentralization emphasizes categorical location: suburbia is a spatially undifferentiated conglomerate of communities outside central cities. Having no internal spatial reference and further muddled by the ebb and flow of central city annexations and defensive suburban incorporations, cross-sectional and longitudinal comparisons based solely on decentralization are misleading.[17] The sec-

[15] B.J.L. Berry, "The Counterurbanization Process: How General?" in *Human Settlement Systems*, ed. N.M. Hansen (Cambridge, Mass.: Ballinger, 1978), ch. 2; B.J.L. Berry, ed., *Urbanization and Counterurbanization*, Urban Affairs Annual Review no. 11 (Newbury Park, Calif.: Sage Publications, 1976); F.J. Coppa, "Cities and Suburbs in Europe and the United States," in *Suburbia: The American Dream and Dilemma*, ed. P.C. Dolce (Garden City: Anchor Books, 1976), pp. 167-191; B. de Borger, "Urban Population Density Functions: Some Belgian Evidence," *Annals of Regional Science* 8.3 (1979): 15-24; Goldberg and Mercer, *The Myth of the North American City*, pp. 151-154; P. Gordon, "Deconcentration without a 'Clean Break,'" *Environment and Planning A* 11 (1979): 281-290; Jackson, *Crabgrass Frontier*, pp. 6-10; D.R. Vining and T. Kontuly, "Population Dispersal from Major Metropolitan Regions: An International Comparison," *International Regional Science Review* 3 (1978): 49-73; D.R. Vining and A. Strauss, "A Demonstration That the Current Deconcentration of Population in the U.S. Is a Clean Break with the Past," *Environment and Planning A* 9 (1977): 751-758; A.J. Fielding, "Counterurbanization," in *Population Geography: Progress and Prospect*, ed. M Pacione (London: Croom Helm, 1986), ch. 8.

[16] Also called political and ecological suburbanization, respectively, the measures describe distinctly different patterns of population distribution within metropolitan areas. However different in their conceptual meanings, as a practical matter of data collection they often rely on the same set of census tract data. The measure of ecological suburbanization provides population density estimates along a smoothly declining gradient; as such it cannot separate suburb from central city. For a discussion of the two measures see A.M. Guest, "Population Suburbanization in American Metropolitan Areas, 1940-1970," *Geographical Analysis* 7 (1975): 267-283; J.D. Kasarda and G.V. Redfearn, "Differential Patterns of City and Suburban Growth in the United States," *Journal of Urban History* 2 (1975): 43-66; E.S. Mills, "Urban Densities," *Urban Studies* 7 (1970): 5-20; J.S. Adams, "Residential Structure of Midwestern Cities," *Annals of the Association of American Geographers* 60 (1970): 37-62; B.J.L. Berry et al., "Urban Population Densities: Structure and Change," *Geographical Review* 53 (1963): 389-405; L.F. Schnore, "The Timing of Metropolitan Decentralization," *Journal of the American Institute of Planners* 25 (1959): 200-206.

[17] K.T. Jackson, "Urban Deconcentration in the Nineteenth Century: A Statistical Inquiry," in *The New Urban History*, ed. L.F. Schnore (Princeton: Princeton University Press), ch. 3. Although Jackson uses the same term to measure the two types of suburbanization, he empha-

ond measure is a function of distance from the city center and sees residential density differentials between a metropolitan center and its periphery diminish. Deconcentration emphasizes relative location: suburbia forms a string of points along a spatially anchored continuum that allows cross-sectional profiles and longitudinal analyses. Suburbanization, as process rather than place, is a matter of gradual change and transition.

Research along these lines has found that today most Americans reside in suburbs, that during the last forty years population densities have declined in suburbs closest to the central city, and that in general the population density gradient for metropolitan areas has been declining.[18] All this points to continued suburban growth and expansion at the periphery accompanied by a noticeable loss of population at and near the center, and thus satisfies the standard criteria of suburbanization.[19]

On Excluding Environmental Factors

Inasmuch as population density has been a widely accepted measure of suburbanization and urban form, natural environmental factors have largely been discounted as additional diagnostic variables in interpreting the geographic patterns of suburban growth and residential location.[20] The dec-

---

sizes the crucial difference by distinguishing between place and process. That occasional conceptual confusion can lead to erroneous and ambiguous conclusions is illustrated by P. Gober and M. Behr, "Central Cities and Suburbs as Distinct Places: Myth or Fact?" *Economic Geography* 58 (1982): 371-385.

[18] U.S. Bureau of the Census, *1980 General Population Characteristics: United States Summary* (Washington, D.C.: U.S. Government Printing Office, 1982); P. Hall, "Decentralization without End? A Re-evaluation," in *The Expanding City*, ed. J. Patten (London: Academic Press, 1983), pp. 125-155; Guest, "Population Suburbanization"; B. Edmonston, *Population Distribution in American Cities* (Lexington, Mass.: Lexington Books, 1975); B. Edmonston, M.A. Goldberg, and J. Mercer, "Urban Form in Canada and the United States: An Examination of Urban Density Gradients," *Urban Studies* 22 (1985): 209-217.

[19] In offering a sustained historical perspective, Ebner suggests the replacement of such conventional terms as "city," "suburb," and "metropolis" with the single phrase "urban population deconcentration," calling it a demographic process providing "a causal explanation encompassing the vast sweep of American urban history since the beginning of the nineteenth century" (p. 368). For a complete discussion see M.H. Ebner, "Re-reading Suburban America: Urban Population Deconcentration, 1810-1980," *American Quarterly* 37 (1985): 368-381.

[20] Agnew, Mercer, and Sopher, introduction to *The City*, pp. 9-21; R.J. Johnston, *The American Urban System* (New York: St. Martins, 1982); R.J. Johnston, *City and Society: An Outline for Urban Geography* (London: Hutchinson, 1984); H.M. Mayer, "A Survey of Urban Geography," in *The Study of Urbanization*, ed. P.M. Hauser and L.F. Schnore (New York: Wiley, 1965), ch. 3; R. Palm, *The Geography of American Cities* (New York: Oxford University Press, 1981).

laration that "cities (are) systems within systems of cities"[21] asked urban geographers to integrate both levels of research and, at the same time, held out the hope that by studying smaller parts of the system one might learn about the whole and vice versa. Twenty-five years have passed and still there is "very little understanding of how to put these different patterns together in more general models."[22] As urban geographic research continued to proceed on two separate levels, it formalized a split view of the potential significance of natural environmental factors.

At the national, interurban level, researchers found environmental conditions to have had a noticeable, if unpredictable, influence on the distribution of urban populations. Physiographic factors were routinely invoked to describe and explain the location of cities and the pattern of early urban growth. City biographies and developmental models of regional and national city systems have focused on specific natural features that, as resources, endowed locations with initial competitive advantages in terms of production, distribution, and consumption.[23] In addition to the standard list of economic, social, demographic, and behavioral variables, the influence of numerous environmental factors as amenities has been tested repeatedly in studies that examine the national pattern of population change and the direction and volume of recent regional and interstate migration and in studies that explore the role of residential preferences in migration behavior.[24]

---

[21] B.J.L. Berry, "Cities as Systems within Systems of Cities," in *Regional Development and Planning*, ed. J. Friedman and W. Alonso (Cambridge: Massachusetts Institute of Technology Press, 1964), ch. 4.

[22] Berry, "Cities as Systems," p. 132.

[23] J.R. Borchert, "American Metropolitan Evolution," *Geographical Review* 57 (1967): 301-322; E.S. Dunn, Jr., *The Development of the U.S. Urban System*, vol. 1 (Baltimore: Resources for the Future, 1983); B.J.L. Berry and E. Neils, "Location, Size, and Shape of Cities Influenced by Environmental Factors: The Urban Environment Writ Large," in *The Quality of the Urban Environment*, ed. H.S. Perloff (Washington, D.C.: Resources for the Future, 1969), ch. 8.

[24] P.E. Graves, "Migration and Climate," *Journal of Regional Science* 20 (1980): 227-237; R.L. Morrill, "Bases for Peripheral Urban Growth," in *Impact of Urbanization and Industrialization on the Landscape*, ed. D.R. Deskins, University of Michigan Department of Geography, Michigan Geographical Publication no. 25 (Ann Arbor: University of Michigan Department of Geography, 1980), pp. 95-117; P.E. Graves and J. Regulska, "Amenities and Migration over the Life-Cycle," in *The Economics of Urban Amenities*, ed. D.B. Diamond and G.S. Tolley (New York: Academic Press, 1982), ch. 10; C.F. Mueller, *The Economics of Labor Migration: A Behavioral Analysis* (New York: Academic Press, 1982); R.J. Cebula, *Determinants of Human Migration* (Lexington, Mass.: Lexington Books, 1979); M.J. Greenwood, "Human Migration: Theory, Models, and Empirical Studies," *Journal of Regional Science* 25 (1985): 521-544; G.R. Hovinen, "Leapfrog Developments in Lancaster County: A Study of Residents' Perceptions and Attitudes," *Professional Geographer* 24 (1977): 194-199; J.D. Williams and D.B. McMillan, "Migration Decision Making among Nonmetropolitan-bound Migrants," in *New*

At the local, intraurban level natural landscape factors, when not completely disregarded, have mostly been discussed in rather general and ambiguous terms.[25] Physiographic features distort the ideal shape of cities and influence the direction of territorial expansion; however, their potential effects on the intensity of land use receive little attention, except where there exist obvious locational advantages concerning the transfer of goods and general interurban access. Why consider natural landscape characteristics within urban areas when entire cities are platted according to the orthogonal geometry of a gridiron?[26] Abstract, two-dimensional space provides "a powerful, practical, and easily manipulated framework to organize people and resources" and thus "increases production and consumption."[27]

More often than not, if considered at all, environmental conditions are viewed as engineering problems.[28] Technology conquers, even obliterates topography and helps in the construction of a geometric, nearly

---

Directions in Rural-Urban Migration, ed. D.L. Brown and J.M. Wardwell (New York: Academic Press, 1980), ch. 8; A.J. Sofranko and F.C. Fliegel, "Neglected Components of Rural Population Growth," *Growth and Change* 14 (1983): 42-49.

[25] In studies of residential neighborhoods natural environmental factors, if at all considered, are subsumed by the physical and locational characteristics of the urban, built environment. The dimensions of residential environmental quality include primarily neighborhood and dwelling characteristics (e.g., traffic, noise, safety, age of dwelling, number of bathrooms, central air/heat). See V. Preston, "A Multidimensional Scaling Analysis of Individual Differences in Residential Area Evaluation," *Geografiska Annaler B* 64 (1982): 17-26; P. Weichhart, "Assessment of the Natural Environment: A Determinant of Residential Preferences," *Urban Ecology* 7 (1982/83): 325-342; W. Michelson, *Environmental Choice, Human Behavior, and Residential Satisfaction* (New York: Oxford University Press, 1977), pp. 113-180.

[26] P.F. Lewis, "Small Town in Pennsylvania," *Annals of the Association of American Geographers* 62 (1972): 323-351; Jackson, *Crabgrass Frontier*, pp. 73-76; J.W. Reps, *Cities in the American West: A History of Frontier Urban Planning* (Princeton: Princeton University Press, 1979); J.W. Reps, *The Forgotten Frontier: Urban Planning in the American West* (Columbia: University of Missouri Press, 1981).

[27] R.D. Sack, *Human Territoriality: Its Theory and History* (Cambridge: University of Cambridge Press, 1986), p. 218. Also see D.M. Gordon, "Capitalist Development and the History of American Cities," in *Marxism and the Metropolis*, ed. W.T. Tabb and L. Sawers (Oxford: Oxford University Press, 1980), pp. 21-53; R.A. Walker, "The Transformation of Urban Structure in the Nineteenth Century and the Beginnings of Suburbanization," in *Urbanization and Conflict in Market Societies*, ed. K.R. Cox (Chicago: Marouffa, 1978), ch. 8. A less emphatic, more straightforward view is given by D. Stanislawski, "The Origin and Spread of the Grid-Pattern Town," *Geographical Review* 36 (1946): 105-120.

[28] An early exception to the tendency to classify residential sites in terms more familiar to a structural engineer is the GASP scheme of urban morphology: growth, accessibility, site, and persistence are identified as the basic categories of structural determinants. B. Duncan, "Variables in Urban Morphology," in *Urban Sociology*, ed. E.W. Burgess and D.J. Bogue (Chicago: University of Chicago Press, 1967), ch. 1.

frictionless city instead of the geomorphic city designed with nature.[29] Adverse physiographic features within urban areas are short-term nuisances or obstacles that will be overcome by mitigation or removal.[30] Why consider such conditions as the pollution of air, soil, and water as long as they remain absent or are ephemeral and do not exceed average levels of tolerance and safety? When urban activities produce public hazards and other nonmarketable, locational disamenities that as negative externalities discourage growth and diminish the value of urban land, only then do environmental factors directly enter geographic analyses of internal urban form.[31] That the mere absence of adverse environmental conditions has been taken to represent positive externalities or amenities reveals a certain reluctance to modify existing analytical models, since it requires neither identification nor measurement of new environmental variables.

Despite the realization that environmental factors have had a noticeable, if variable and unpredictable, influence on population distribution at the national level, recognition of this influence at the local and individual

---

[29] "Friction of space" is overcome by transportation. Transportation costs and site rent are the chief components of the "costs of friction." Accessibility thus becomes a good, just like food and clothes. So goes the argument of R.M. Haig, "Toward an Understanding of Metropolis, Parts I and II," *Quarterly Journal of Economics* 40 (1926): 179-208 and 402-434. Urban design that complements and takes advantage of the natural environment and conditions is advocated by I. McHarg, *Design with Nature* (Garden City, N.Y.: Natural History Press, 1969).

[30] Consider, for example, Chicago's own "eighth wonder of the world": reversing the flow of the Chicago River channeled wastes into the Mississippi River basin instead of into Lake Michigan, preventing contamination of the city's major source of drinking water. In the end, the construction of the Chicago Sanitary and Ship Canal had other beneficial spillover effects such as flood control, improved ship and barge transportation, and provision of new sites for industrial relocation and development. See H.M. Mayer and R.C. Wade, *Chicago: Growth of a Metropolis* (Chicago: University of Chicago Press, 1969), p. 274. For more detailed studies on the historic development along the canal see M.P. Conzen and M.J. Morales, eds., *Settling the Upper Illinois Valley: Patterns of Change in the I&M Canal Corridor, 1830-1900*, University of Chicago Committee on Geographical Studies, Studies on the I&M Canal Corridor no. 3 (Chicago: University of Chicago Committee on Geographical Studies, 1989).

[31] B.J.L. Berry et al., *Land Use, Urban Form, and Environmental Quality*, University of Chicago Department of Geography Research Paper no. 155 (Chicago: University of Chicago, Department of Geography, 1974); B.J.L. Berry et al., *The Social Burdens of Environmental Pollution: A Comparative Metropolitan Data Source* (Cambridge, Mass.: Ballinger, 1977); B.J.L. Berry, and F.E. Horton, *Urban Environmental Management: Planning for Pollution Control* (Englewood Cliffs: Prentice Hall, 1974); J.A. Schmid, "The Environmental Impact of Urbanization," in *Perspectives on Environment*, ed. M.W. Mikesell and I.R. Manners (Washington, D.C.: Association of American Geographers Commission on College Geography, 1974); ch. 8; H.S. Perloff, "A Framework for Dealing with the Urban Environment: Introductory Statement," in *The Quality of the Urban Environment*, ed. H.S. Perloff (Baltimore: Johns Hopkins Press, 1969), ch. 1; D.M. Elsom, "Pollution," in *Progress in Urban Geography*, ed. M. Pacione (London: Croom Helm, 1983), ch. 10.

level has been further hampered by the persistent lack of agreement concerning compatible standards for the identification and measurement of such factors. There is general uncertainty concerning their relative order of significance or levels of interdependence.[32] A specific environmental condition or landscape feature may be highly preferred and valued in one location and yet be completely overlooked or considered commonplace in another. Additional complications arise as the definition of valued landscapes and landscape components may be intimately tied to diverse public policy as well as to less formal societal and personal goals of achievement and satisfaction, all of which may vary not only from place to place, but also through time.[33]

Inasmuch as the selection of environmental variables depends on data availability and compatibility, the decision to include such variables is ultimately based on a study's geographic and temporal span. Climatic variables have been the favorite source for regional and national comparison. They are the most easily obtained environmental data with unsurpassed geographic and temporal coverage and can provide the basis for air quality simulation models of selected metropolitan regions.[34] Composite, objective indices of quality of life have been computed to classify and rank metropolitan areas based on differences in weather conditions, pollution levels, outdoor recreational facilities, and even urban-suburban population densities.[35] By and large, studies of this kind have had to rely on environmental proxy measures that generated interval or ratio scales when observed qualitative differences proved insufficient or were unavailable. However, the continued lack of interest in examining the influence of natural environ-

---

[32] S.L. Cutter, *Rating Places: A Geographer's View on the Quality of Life*, Association of American Geographers Resource Publications in Geography (Washington, D.C.: Association of American Geographers, 1985); E.H. Zube, "Rating Everyday Rural Landscapes of the Northeastern United States," *Landscape Architecture* 63 (1973): 370-375; E.H. Zube et al., *Landscape Assessment: Values, Perceptions, and Resources* (Stroudsburg, Pa.: Dowden, Hutchinson, and Ross, 1975).

[33] E.H. Zube, *Environmental Evaluation: Perception and Public Policy* (Monterey, Calif.: Brooks-Cole, 1980).

[34] I. Hoch and J. Drake, "Wages, Climate, and the Quality of Life," *Journal of Environmental Economics and Management* 1 (1974): 284-295; R.S. Bednarz, *The Effect of Air Pollution on Property Value in Chicago*, University of Chicago Department of Geography Research Paper no. 166 (Chicago: University of Chicago Department of Geography, 1975); G.Y. Lin, "Simple Markov Chain Model of Smog Probability in the South Coast Air Basin of California," *Professional Geographer* 33 (1981): 228-236.

[35] B.C. Liu, *Quality of Life Indicators in U.S. Metropolitan Areas: A Statistical Analysis* (New York: Praeger, 1976); Berry et al., *The Social Burdens*; I. Hoch, "Variations in the Quality of Urban Life among Cities and Regions," in *Public Economics and the Quality of Life*, ed. L. Wingo and A. Evans (Baltimore: Johns Hopkins University Press, Resources for the Future and Centre for Environmental Studies, 1977), ch. 5.

mental factors on urban form and structure is not solely the result of the practical issue of data collection and measurement within a national or regional context. It is as much the result of a profound reorientation, even redefinition, of urban geographic research, which through theoretical, often contentious, debate has sought to identify and sanction proper research topics and explanation.

Choices of Ideology and Methodology

Social class and demographic composition, ethnicity and race, residential mobility and job location, employment and income, political control and fragmentation, provision of public and commercial services, land use and land rent, transportation and daily commuting: these have been the major themes for studying the internal structure and growth of urban areas.[36] Each of these themes opens a distinctly different window on suburbanization as it allows researchers not only to set suburban communities apart from their central city, but also to distinguish among them. How these themes tie together is not clearly understood. How their principal elements influence one another or the general distribution of population within metropolitan areas is only incompletely documented. There is obviously little, if any, provision to consider the role of natural landscape features.

Around the above-mentioned themes of urban structure and organization has emerged a set of approaches that advocate competing process-oriented theories of greatly divergent philosophy and methodology. The behavioral-perceptual, institutional-managerial, and Marxist political approaches ostensibly evolved in opposition to the positivist orthodoxy and reductionism of neoclassical economics and human ecology.[37] So radically different in conceptualization and application, not one of the three alternatives has yielded either a general theory of urban structure or an acceptable

---

[36] Palm, *The Geography of American Cities*. Also see the Annual Status Reports entitled "Urban Geography: City Structure" in *Progress in Human Geography*, 1-7, authored by R.J. Johnston (1977-1980) and R. Palm (1981-1983).

[37] M.T. Cadwallader, "Urban Geography and Social Theory," *Urban Geography* 9 (1988): 227-251. Commentaries by W.A.V. Clark, E. Sheppard, and J.S. Duncan follow on pp. 252-268. Other researchers have preferred to distinguish between positivist, behavioral, humanist, and structuralist traditions; see P. Jackson and S.J. Smith, *Exploring Social Geography* (London: Allen and Unwin, 1984); and M. Gottdiener, "Understanding Metropolitan Deconcentration: A Clash of Paradigms," *Social Science Quarterly* 64 (1983): 227-246. A less radical view foresees pluralism within the subfield to increase as quantitative (i.e., positivist) human geography seeks to accommodate the challenges of humanistic and radical ideologies. See R.J. Johnston, "Ideology and Quantitative Human Geography in the English-Speaking World," in *European Progress in Spatial Analysis*, ed. R.J. Bennett (London: Pion, 1981), ch. 2.

middle ground of synthesis and explanation, a failure they share with the positivist tradition.[38]

Of the three, only the behavioral approach has recognized the possible importance of the natural environment when interpreting the processes and patterns of residential relocation and locational preference.[39] The study of proclaimed residential preferences has contributed to a better understanding of actual and planned residential mobility of individuals and small groups. Attitude-discrepant behavior, brought about by a variety of individual or institutional constraints that may be at odds with declared environmental images, represents a concept that fills some of the gaps between model and data.[40] Unfortunately, there is little consistency concerning the identification and relative ranking of determinants; and there remains substantial uncertainty about the decision-and-search process itself.[41]

Having either rejected or discredited the positivist assertion that the study of aggregated spatial patterns permits inferences about economic, social, and political processes, researchers have sought to comprehend the societal structures and their underlying economic and political linkages. By exploring the motives, attitudes, perceptions, and aspirations of individuals and groups and by dissecting the allocation of political and economic power, they have hoped to learn more about the development and inner workings

---

[38] Much of the criticism aimed at positivist methodology and attendant perspectives identified the desire for formulating universally applicable laws as one of the more serious shortcomings of this investigative tradition. Even though Marxist social theory is driven by a similarly sweeping claim, the scope of current inquiry is generally less expansive and favors the interpretations of conditions and circumstances on their own terms within a particular spatial and temporal setting.

[39] More often than not neighborhood characteristics and dwelling attributes define the extent to which environmental dimensions have entered behavioral models of intraurban mobility and residential preferences. Natural environmental characteristics, if included at all, are usually couched in rather ambiguous terms or lumped together as a single categorical variable. See L.Y. Mudrak, "Sensory Mapping and Preferences for Urban Nature," *Landscape Research* 7 (1982): 2-8; L.Y. Mudrak, "Urban Residents' Landscape Preferences: A Method for Their Assessment," *Urban Ecology* 7 (1983): 91-123; J. Drewnowski, *On Measuring and Planning the Quality of Life*, Publications of the Institute of Social Studies no. 11 (The Hague: Mouton, 1974); M. Pacione, "Revealed Preferences and Residential Environmental Quality," in *Quality of Life and Human Welfare*, ed. M. Pacione and G. Gordon (Norwich, U.K.: Geog Books, 1984), ch. 5; M.T. Cadwallader, "Neighborhood Evaluation in Residential Mobility," *Environment and Planning A* 11 (1979): 393-401.

[40] J.M. Desbarats, "Spatial Choice and Constraints on Behavior," *Annals of the Association of American Geographers* 73 (1983): 340-357; A.B. Shlay, "Taking Apart the American Dream: The Influence of Income and Family Composition on Residential Evaluations," *Urban Studies* 23 (1986): 253-270.

[41] M.T. Cadwallader, "Migration and Intra-Urban Mobility," in *Population Geography: Progress and Prospect*, ed. M. Pacione (London: Croom Helm, 1986), ch. 9.

of urban complexes. The study of urban forms would remain needlessly reductionist and simplistic in perspective and interpretation unless it sought firm footing within a broader sociopolitical and cultural environment.[42] As cultural values and value systems permeate and mold the interlocking layers of social organization, political institutions, and economic systems, the city becomes more than simply a container of culture; rather, the city is both sediment and agent of culture. Urban form is transformed only when a counter-culture deposes the previously dominant, "hegemonic culture."[43] Cultural competition, articulated by cultural values, replaces locational competition, driven by rational economic behavior, and becomes a major theme in the study of urban growth and development.

The concept of cultural context promises a perspective that is richer in detail and more incisive in interpretation than any of the four research traditions. The practical nature of everyday life should be the new nucleus for explanations and interpretations of urban form, not the abstract calculus of economic production and consumption that dominates positivist interpretations and underlies to varying degrees the other approaches.[44] Such a reconstructed research perspective would cling to the tenets of social theory.[45] The notion that spatial structures such as cities are as much the outcome as they are the conveyor of ideas, beliefs, and values remains ambiguous in concept and does not offer a clear choice of subject. To speak of a "convergence of time-space analysis" and the "reflexive impact of space on society" never diminishes the singular importance of social process as central object of inquiry.[46] Clearly, the identification, let alone the analysis, of spatial patterns (e.g., urban form) is of secondary importance.

---

[42] S.J. Lewandowski, "The Built Environment and Cultural Symbolism in Post-Colonial Madras," in *The City in Cultural Context*, ed. J.A. Agnew, J. Mercer, and D.E. Sopher (Boston: Allen and Unwin, 1984), p. 252.

[43] Goldberg and Mercer, *The Myth of the North American City*, pp. 5-6.

[44] Agnew, Mercer, and Sopher, introduction to *The City*, p. 3. The same programmatic conclusion echoes in the following statement: "Social theory is concerned with the illumination of the concrete process of everyday life. Human geography, therefore, can be construed as that part of social theory which focuses on the spatial patterns and processes which underlie the structures and appearances of everyday life." M. Dear, "The Postmodern Challenge: Reconstructing Human Geography," *Transactions of the Institute of British Geographers* N.S. 13 (1988): 267; also see J.S. Duncan and N.D. Duncan, "A Cultural Analysis of Urban Residential Landscapes in North America: The Case of the Anglophile Elite," in *The City in Cultural Context*, ed. Agnew, Mercer, and Sopher, ch. 12.

[45] As to the difference between "cultural" and "social," there exists a curious ambivalence in usage. The terms are often used interchangeably. When the proposals by the Duncans and Dear are compared, the arguments of the former appear less strident in tone and less ideologically informed than those of the latter.

[46] Dear, "The Postmodern Challenge," pp. 267 and 269.

It is not culture's imprint on the landscape that matters as physical evidence; empirical evidence detailing the cultural context, the social structures as they constrain and free individuals to act, has priority.[47] The relevance of spatial attributes such as natural landscape features or population density is entirely dependent on their symbolic meaning and the extent to which they accompany and articulate social and individual action.[48] The seams that bind and separate social structure and its institutional and human agents are extremely fluid boundaries. This has the effect of further blurring their respective correspondence with spatial patterns that have been or are being formed as a result. Spatial elements and structures serve as accessories in making complete the self-image and identity of groups and individuals. As self-image and identity change and conflicts among groups arise, accessories may be discarded, retained, or given entirely new meaning. As epiphenomenon the physical urban landscape is virtually removed from the study of urban geography thus oriented.

Whether patterns of population density within urban areas could be addressed in similar fashion seems unlikely. Such aggregate patterns would have to be dissolved into separate social areas, each having a specific inventory of symbolic spatial accessories. At the macro- level, however, population density gradients have served as a useful empirical measure when examined in light of broadly defined cultural traits of national character. A comparison of metropolitan areas in Canada and the United States suggests that observed differences in gradients are an expression of deep-seated differences in social, political, and economic values and institutions.[49] The American preference for suburban, quasi-rural living was listed as one of the factors that favored a more spread-out, less compact urban form in the United States.[50]

---

[47] This single-minded orientation is clearly expressed by the following: "[W]e will argue that culture is not an explanatory variable. On the contrary, it is what is to be explained or, less ambitiously, commented upon. Its complexity must be prised apart; one must not only discover the origins of various cultural elements, but also show what these elements mean to people, their relationship to group and individual identity, to what ends they are put, and the manner in which people struggle or fail to struggle to maintain or change them." Duncan and Duncan, "A Cultural Analysis," p. 256.

[48] Carried to the extreme, group analytical techniques developed in psychotherapy have been used to reveal environmental values and open-space preferences. See J. Burgess et al., "Exploring Environmental Values through the Medium of Small Groups, Part 2: Illustrations of a Group at Work," *Environment and Planning A* 20 (1988): 457-476.

[49] Edmonston, Goldberg, and Mercer, "Urban Form in Canada and the United States"; Goldberg and Mercer, *The Myth of the North American City*, pp. 139-152.

[50] Goldberg and Mercer, *The Myth of the North American City*, p. 147.

As one of the most enduring spatial expressions of urban form, population density gradients have been criticized for being overly simplistic in their theoretical arguments supporting the original negative exponential specification. Density gradients belong to a collection of spatial models invariably considered anachronistic, empty constructs devoid of any analytical or diagnostic power, since their buttressing, inferred economic and social processes, as well as some of the necessary preconditions such as the continued influx and growth of population at the center, have either changed or ceased to exist.[51] Their spatial representations display the familiar geometries of concentric rings, sectoral wedges, and linear gradients as cross-sectional or one-dimensional equivalents of the first two. Since the observed persistence of certain patterns, such as concentricity and sectorality, appears to be out of step with the recent shifts in societal structure, it has been argued that pattern be separated from process and that the two be studied independently.[52]

These models, in their own way, contributed to the general neglect of natural environmental features as potentially important factors. Clearly, the assumption that all urban activity is focused on a single center located on a featureless plain needs to be reexamined in view of the increased scale and fragmentation of modern metropolitan complexes.[53] Instead of categorically declaring any of these models obsolete, this study takes the position that by adding new variables these models can still fulfill a useful function in exploring the spatial structure of metropolitan areas.

Theoretical incompatibility notwithstanding, all of these approaches have remained largely independent because of seemingly insurmountable differences in scale. Competing methodologies built upon distinct philosophies have been unable to integrate the various scales of study (macro-, meso-, and micro-), compounding the problem of reconciling form with function, structure with agent, pattern with process. Furthermore, differences in data collection (empirical-subjective vs. abstract-objective) still pre-

---

[51] A complete rejection of the underlying factors and processes is proposed by B. Marchand, *The Emergence of Los Angeles* (London: Pion, 1986). A less polemical discussion is presented by W. Alonso, "The Population Factor and Urban Structure," in *The Prospective City*, ed. A.P. Solomon (Cambridge: Massachusetts Institute of Technology Press, 1980), ch. 1.

[52] B. Marchand, "Urban Growth Models Revisited: Cities as Self-Organizing Systems," *Environment and Planning A* 16 (1984): 949-964.

[53] P.G. Goheen, "Interpreting the American City," *Geographical Review* 64 (1974): 362-384; D.A. Griffith, "Evaluating the Transformation from a Monocentric to a Polycentric City," *Professional Geographer* 33 (1981): 189-196; A. Kutay, "Technological Change and Spatial Transformation in an Information Society, Part I: A Structural Model of Transition in an Urban System"; and "Part II: The Influence of New Information Technology on an Urban System," *Environment and Planning A* 20 (1988): 569-593 and 707-718.

vent the integration needed for a general, overarching theory.[54] Polemics against the lifeless reductionism of positivist methodology and its sharp focus on spatial patterns have not been followed by resolutely documented research.[55] Whether humanist, structuralist, or Marxist in orientation, none of these approaches has presented linkages between pattern and process that are any stronger, more convincing, or more encompassing than those presented by the maligned positivist tradition.[56]

The notion that additional insights into urban structure may be reached by including natural environmental features is not new. Whereas analysis at the macro- scale has readily incorporated natural environmental variables, it still is uncertain in what manner and to what extent they can be addressed at the meso- scale. To employ positivist methodology, as this study does, by no means dismisses the relative significance or relevance of other approaches. However, the positivist framework is better suited for the kind of problem examined in this study, particularly as it offers a set of spatial models that can be modified for comparative purposes. Analysis of the spatial association between suburban population densities and natural landscape features does not seem readily to fit any of the other approaches, as it runs counter to their respective research agendas, both in terms of topical interests and ideological orientation.[57] Yet by offering such an analysis, this study focuses on a set of formal geographic elements without which a basic understanding of suburban settlement in this country remains incomplete and flawed.

---

[54] Comment by W.A.V. Clark on M.T. Cadwallader, "Urban Geography and Social Theory," *Urban Geography* 9 (1988): 252-254.

[55] R.I. Woods, "Theory and Methodology in Population Geography," in *Population Geography: Progress and Prospect*, ed. M. Pacione (London: Croom Helm, 1986), ch. 2.

[56] Dear's diagnosis of geography as a discipline in depressing disarray is in part based on the notion that "[t]he only common ground which presently unites some groups of geographers is a predilection toward positivism to the exclusion of all other approaches" (p. 265). Later, he softens his stand somewhat by accepting the principal relativism of human knowledge; however, he is not prepared to embrace the "Anything Goes" school of philosophy (p. 272). Instead he argues for human geography to adopt social theory as its central paradigm, thereby advocating a step that he accuses "rising stars" and "embittered elders" of having taken to the detriment of the discipline's identity and unity: namely, that of looking outside geography in order to find focus and cohesion (p. 256). Dear, "The Postmodern Challenge."

[57] H. Leitner, "Urban Geography: Undercurrents of Change," *Progress in Human Geography* 11 (1987): 134-146.

Chapter 3

CONCEPTS OF THE NATURAL ENVIRONMENT AS AMENITY IN AN URBAN SOCIETY

> *[A]n amenity can be defined in many ways. Some have linked it with qualities of desirability which lead to enhanced economic value of properties or which exert a lure to potential immigrants; some define it as any phenomenon which results in a pleasurable experience to those who are exposed to it; while others suggest that it is any comfort or convenience beyond the level of life's necessity. Some include man-made facilities in their definition while others confine their use of the term to certain natural features of the environment.*
> —A.A. Atkisson and I.M. Robinson, "Amenity Resources for Urban Living"

Warm-dry climates, mountains, beaches, and areas that support a wide range of outdoor recreational activities were singled out as major natural amenities when in 1954 Ullman called attention to their increased significance as locational factors of urban and regional population growth.[1] Before him Hurd and Hoyt had already alluded to the general importance of pleasant physiographic conditions as they affected the distribution of people, land uses, and values within cities.[2] Other writers since then have fleshed out the idea of amenities as influential and independent variables affecting the allocation of people and economic activities.[3] Accordingly, environ-

---

[1] E.L. Ullman, "Amenities as a Factor in Regional Growth," *Geographical Review* 44 (1954): 119-132.

[2] R.M. Hurd, *Principles of City Land Values* (New York: Record and Guide, 1924); H. Hoyt, *The Structure and Growth of Residential Neighborhoods in American Cities* (Washington, D.C.: U.S. Government Printing Office, 1939).

[3] H.S. Perloff et al., *Regions, Resources, and Economic Growth* (Baltimore: Johns Hopkins Press, Resources for the Future, 1960), pp. 471-475; H.S. Perloff and L. Wingo, Jr., "Natural Resource Endowment and Regional Economic Growth," in *Regional Development and Planning*, ed. J. Friedman and W. Alonso (Cambridge: Massachusetts Institute of Technology

mental amenities are crucial locational attributes because consumption motives and preferences for qualitative improvements of nonmarketable goods and all kinds of externalities have become more important than conventionally defined goods, commodities, and input factors of production. Amenities that count are viewed as one more expression of imminent structural changes transforming developed space-economies. In their purest form amenities are the classic nonproduced public goods and as such provide extreme externalities.[4] They are indivisible and cannot be transferred, though they may be replicable. As location-fixed goods they are consumed in place and consumption is automatic.

Even though the general concept of environmental amenity seems to have been accepted, the definition of the specific role amenities play depends on the interpretative framework within which a study operates. The economist's interpretation, sketchily outlined in the previous paragraph, provides but one view among many. This chapter discusses several geographic interpretative frameworks and compares them in terms of their capacity to identify specific amenities, their interrelationships, and their impact on the distribution of urban populations. This is not an attempt at a comprehensive survey of the topic; it merely serves to illustrate the divergent conceptual and operational approaches that have emerged during the last two decades and which have grappled directly or indirectly with the question of how to incorporate the notion of amenity into existing modes of research and interpretation.

Press, 1964), ch. 11; H.S. Perloff, "A Framework for Dealing with the Urban Environment: Introductory Statement," in *The Quality of the Urban Environment*, ed. H.S. Perloff (Baltimore: Johns Hopkins Press, 1969), ch. 1; M.M. Webber, "The Urban Place and the Nonplace Urban Realm," in *Explorations into Urban Structure*, ed. M.M. Webber et al. (Philadelphia: University of Pennsylvania Press, 1964), ch. 2; L. Wingo, "Objective, Subjective, and Collective Dimensions of the Quality of Life," in *Public Economics and the Quality of Life,* ed. L. Wingo and A. Evans (Baltimore: Johns Hopkins University Press, Resources for the Future and Centre for Environmental Studies, 1977), ch. 1; J.A. Burgess, "Selling Places: Environmental Images for the Executive," *Regional Studies* 16 (1982): 1-17; J.F. Richardson, "The Dynamics of American Urban Development," in *Cities in the Twenty-first Century*, ed. G. Gappert and R.V. Knight (Newbury Park, Calif.: Sage Publications, 1982), ch. 2; J.M. Wardwell, "Toward a Theory of Rural-Urban Migration in the Developed World," in *New Directions in Urban-Rural Migration,* ed. D.L. Brown and J.M. Wardwell (London: Academic Press, 1980), ch. 4.

[4] Defined in this fashion an amenity's benefits are available to everyone free of charge; and the use of it by any individual does not reduce the amount available to others. For further discussion see H.S. Perloff's contributions listed in preceding footnote. A recent, rather broad definition is provided by D.B. Diamond and G.S. Tolley, "The Economic Roles of Urban Amenities," in *The Economics of Urban Amenities,* ed. D.B. Diamond and G.S. Tolley (New York: Academic Press, 1982), ch. 1.

Cultural Values and Predispositions

Cultural factors are the ideal understandings circumscribing everyday social, political, and economic processes and behavior.[5] Representing shared images and myths, a series of cultural predispositions underlie and largely define all other facets of collective and individual life and any interpretation thereof. The cultural history of the United States has been shaped by several such predispositions. The impact of each has been variable in time and across space. They are not necessarily complementary and may even be contradictory. Telltale signs, faint or plain, can be detected in the organization of political representation, the choices of religious expression, and the patterns of urban and rural settlement. Zelinsky discerns four such predispositions, calling them (1) individualism and antiauthoritarianism, (2) mobility and change, (3) mechanistic view of the world, and (4) messianic perfectionism.[6]

Individualism is considered the most powerful and pervasive ideal of American culture.[7] Upon it is constructed "the image of a contented and virtuous rural people";[8] an image that promotes the romantic notion of self-sufficiency and rugged individualism embodied by the heroic figure of the yeoman and later the frontier farmer, and, by extension, fuels a persistent antiurban bias.[9] Against this background the diffuse and seemingly uncoordinated pattern of suburban sprawl is nothing but an "uneasy compromise between a rural tropism and the functional imperative of the city."[10] Rooted in this sentiment is the bewildering political fragmentation that pits

---

[5] B. London and W.G. Flanagan, "Comparative Cultural Ecology: A Summary of the Field," in *The City in Comparative Perspective*, ed. J. Walton and L.H. Masotti (New York: Wiley, 1976), pp. 41-66.

[6] W. Zelinsky, *The Cultural Geography of the United States* (Englewood Cliffs: Prentice Hall, 1973). Glacken asks whether "there [is] merely a group of loosely held and perhaps contradictory notions with unexamined predispositions." He suggests that American attitudes toward nature were at first derivative and only later evolved a native formulation with a strong technical and scientific bias. C.J. Glacken, "Reflections on the Man-Nature Theme as a Subject of Study," in *Future Environments of North America*, ed. F.F. Darling and J.P. Milton (Garden City: Natural History Press, 1966).

[7] This theme is given fuller treatment by J.R. Pole, *American Individualism and the Promise of Progress* (Oxford: Clarendon Press, 1980).

[8] Y.F. Tuan, *Topophilia* (Englewood Cliffs: Prentice Hall, 1974), p. 109.

[9] J.K. Hadden and J.J. Barton, "An Image That Will Not Die: Thoughts on the Anti-Urban Ideology," in *The Urbanization of the Suburbs*, ed. L.H. Masotti and J.K. Hadden (Newbury Park, Calif.: Sage Publications, 1973), ch. 3; L. Marx, "The Puzzle of Antiurbanism in Classic American Literature," in *Cities of the Mind*, ed. L. Rodwin and R.M. Hollister (New York: Plenum Press, 1984), ch. 10.

[10] Zelinsky, *The Cultural Geography*, p. 49.

a multitude of suburban governments and jurisdictions against the landlocked, monolithic central city. Not meant to cope with the massive inmigration of central urban residents, the political structure of the outlying rural townships allowed suburban newcomers, developers and residents alike, to create small-scale entities that as homogeneous enclaves protected its denizens from the perceived negative influences of the crowded industrial city.[11] The garden city of the late nineteenth century was a response to deep-seated rural values and aspirations that cherished the individual and despised government authority. Suburb, as settlement type, evolved into a separate cultural category in antithesis to the city.[12]

Mobility and change are intimately tied to the first theme. Within the urban context the process has been couched in terms of invasion and succession, withdrawal and replacement, and the simple need to move elsewhere.[13] The cultural propensity of wanting to move not only involves spatial mobility, but social mobility as well. The move to suburban locations represents one's removal from the central city and one's attempt at living the pastoral ideal. As improvements in transportation and communication technology facilitate spatial mobility and make it affordable for a growing proportion of the population, the potential of realizing individual freedom and independence diminishes.

The last two predispositions of American culture are contradictory forces that contribute to conflicting spatial patterns of settlement. The mechanistic view of the world finds its tangible expression in the rectangular

---

[11] Low population thresholds for municipal incorporation were designed so as not to burden rural, farming residents with unnecessary government functions and overhead. For a time it took as few as 100 residents to incorporate in Illinois or 500 in California. According to the *U.S. Census of Governments* (1982) the Chicago SMSA had a total of 1,194 different local governments. Among the ten largest SMSAs in the nation (ranked by population size) Philadelphia (867) and Houston (622) came in second and third. San Antonio, with only 77 separate local governments, had the lowest number. Almost half of Chicago's total are in Cook County alone, with the remainder nearly evenly distributed among the other five counties. For further discussion see D.J. Zeigler and S.D. Brunn, "Geopolitical Fragmentation and the Pattern of Growth and Need: Defining the Cleavage between Sunbelt and Frostbelt Metropolises," in *The American Metropolitan System*, ed. S.D. Brunn and J.O. Wheeler (New York: Wiley, 1980), ch. 6.

[12] C. Perin, *Everything in Its Place: Social Order Land Use in America* (Princeton: Princeton University Press, 1977); H.C. Binford, *The First Suburbs: Residential Communities on the Boston Periphery, 1815-1860* (Chicago: University of Chicago Press, 1985).

[13] E.W. Burgess, "The Growth of a City: An Introduction to a Research Project," in *The City*, 7th ed., ed. R.E. Park (Chicago: University of Chicago, 1974), pp. 47-62; Marx, "The Puzzle of Antiurbanism"; P.A. Morrison and J.P. Wheeler, "The Image of 'Elsewhere' in the American Tradition of Migration," in *Human Migration*, ed. W.H. McNeill and R.S. Adams (Bloomington: Indiana University Press, 1978); J. Sonnenfeld, "Variable Values in Space and Landscape," *Journal of Social Issues* 22.4 (1966): 71-82.

grids of urban street plans and the townships-range system that with repetitious regularity divides most of the nation's territory. Cities were laid out and platted with total disregard for topography or other natural obstacles.[14] Intraurban mobility was enhanced, making the transfer of people and goods more efficient and potentially equitable. This straightjacket of spatial uniformity is counterbalanced by the irregular street plan of the garden suburb. Lacking defined points of origin, a garden suburb's streets are curvilinear, even if they do not follow variations in topography and hydrography.[15] The pastoral vision of a secular landscape recognizes the intrinsic qualities of environmental amenities and may well be traced to a preponderance of Anglophile landscape tastes.[16] While many Americans profess their aspiration of living in a rural Eden, only a few can afford to create the appearance of such an ideal. Zelinsky describes it as "a humanized, parklike blending of the best of wild nature with the blessings of civilization, neither wilderness nor city, but a blissful balance between primitive and the urbane."[17] In the extreme, the American suburb becomes "heaven on earth"; the move to the suburbs is a process of moral significance, "a believed-in social evolution, inevitable and right."[18]

Instead of looking to identify long-lasting cultural traits, Vance focuses on the material manifestations of American culture; the single-family detached house and the city in the garden, the urban version of isolated farmsteads, are the preferred residential structures.[19] He distinguishes three major components of the U.S. urban settlement pattern: central city, suburbia, and exurbia/Arcadia. The last is favored by "pastoralists" and "primitivists" and incorporates the dominant spatial metaphors of American culture: the garden, the frontier, the West, and the wilderness.[20] Central city and suburb differ from one another in that the suburb expresses a generational shift in residential location. Foreign immigrants express their successful assimilation into native culture by moving to the suburb.

---

[14] P.F. Lewis, "Small Town in Pennsylvania," *Annals of the Association of American Geographers* 62 (1972): 323-351.

[15] J.G. Fabos et al., *Frederick Law Olmsted, Sr.* (Amherst: University of Massachusetts Press, 1968).

[16] P.J. Hugill, "Home and Class among the American Landed Elite," in *The Power of Place*, ed. J.A. Agnew and J.S. Duncan (London: Unwin and Hyman, 1989), pp. 66-80.

[17] Zelinsky, *The Cultural Geography*, p. 64.

[18] Perin, *Everything in Its Place*, p. 216.

[19] J.E. Vance, "California and the Search for the Ideal," *Annals of the Association of American Geographers* 62 (1972): 185-210.

[20] Tuan, *Topophilia*, pp. 192-197.

Exurbia is located within reach of periodic commuting to the central cities of the eastern megalopolis; Arcadia seeks out locations in beneficent landscapes and is scattered throughout the West. Their essential attribute is a sense of detachment that is defined both geographically and socially. Exurbia symbolizes a spatial compromise between ideal and normative reality, as it is functionally still part of a metropolitan region. Arcadia, on the other hand, needs constantly to redefine its social and spatial distinctiveness. As potentially beneficent landscapes providing environmental amenities have become more accessible, Arcadia's continued existence as a separate American settlement type is in peril.

While research of this type does little to help distinguish specific landscape elements, it unequivocally recognizes and underscores the intrinsic significance that physiographic elements have in defining suburban and exurban locations as the quintessential American settlement type. The description of nature is couched in broad, iconic terms. The assessment of particular natural factors or conditions is deferred in favor of a coarsely woven backdrop against which local conditions may be projected to create interpretative images of varying detail and duration. Cultural predispositions are the relatively permanent components of such images, yet their specific articulation may vary considerably and is ultimately unpredictable.

A number of studies have attempted to gauge personal landscape preferences in order to assess the value and relative importance of specific features or entire landscape ensembles. Although they are not designed to confirm the existence of cultural traits, such preference studies reveal pertinent aspects of the broader cultural context and a person's socioeconomic background. Based on questionnaires and interviews employing either verbal landscape descriptions or visual material, researchers have found that a favorable impression of scenic quality is strongly related to topographic variation, vegetational cover, surface water, and the general diversity and contrast of land-use elements.[21] Overall, Americans appear to prefer park-

---

[21] K.H. Craik, "Appraising the Objectivity of Landscape Dimensions," in *Natural Environment*, ed. J.V. Krutilla (Baltimore: Johns Hopkins University Press, Resources for the Future, 1972), pp. 292-346; D.B. Carruth, "Assessing Scenic Quality," *Landscape* 22 (1977): 31-34; D.L. Linton, "The Assessment of Scenery as a Natural Resource," *Scottish Geographical Magazine* 84 (1969): 219-238; W.G. Hendrix and J.G. Fabos, "Visual Land Compatibility as a Significant Contribution to Visual Landscape Quality," *International Journal of Environmental Studies* 8 (1975): 21-28; R.S. Ulrich, "Visual Landscape Preferences: A Model and Application," *Man-Environment Systems* 7 (1977): 279-293; E.L. Shafer and J.E. Hamilton, "Natural Landscape Preferences: A Predictive Model," *Journal of Leisure Research* 1 (1969): 1-19; N.D. Weinstein, "The Statistical Prediction of Environmental Preferences: Problems of Validity and Application," *Environment and Behavior* 4 (1976): 611-626; J.D. Porteous, "Urban Environmental Aesthetics," in *Environmental Aesthetics: Essays in Interpretation*, ed. B. Sadler and A. Carlson,

like settings with trimmed grass, an absence of tangled underbrush, and scattered shade trees.[22] Such landscapes provide both openness and seclusion as well as a mixture of nonthreatening mystery and comforting familiarity.[23] One author concludes that the color green is nearly uniformly associated with things natural and that it is the apparent orderliness of managed natural landscapes with well-defined edges separating major landscape components that produces strong positive reactions.[24] Both privacy and aesthetically pleasant surroundings seem to be particularly important dimensions of residential satisfaction and have a longstanding tradition in suburban areas.[25]

Have the cultural predispositions suggested by Zelinsky been a uniformly significant influence on the distribution of population in this country? In order to characterize the temporal shifts in the distribution of population, the next section reviews pertinent models of the evolution of the American urban system. Such temporal perspective may shed additional light on recognized stages of urban settlement evolution and thus help to pinpoint observable spatial changes associated with environmental amenities.

University of Victoria Department of Geography, Western Geographical Series no. 20 (Victoria: University of Victoria Department of Geography, 1982), ch. 4.

[22] Speculation on the developmental stages of human landscape preferences has introduced the term "savannah-type landscape," inferring that instinctive biological, rather than learned cultural, preferences shape our perception and evaluation of natural landscape types. See J.D. Balling and J.H. Falk, "Development of Visual Preferences for Natural Environments," *Environment and Behavior* 14 (1982): 5-28. Others suggest that environmental evaluation is subordinate to the concept of humanism in that it addresses the intrinsic meanings of the environment. See J.R. Gold and J. Burgess, eds., *Valued Environments* (London: Allen and Unwin, 1982); and B. Sadler and A. Carlson, "Environmental Aesthetics in Interdisciplinary Perspective," in *Environmental Aesthetics*, ch. 1.

[23] J. Appleton, *The Experience of Landscape* (New York: Wiley, 1975); T.R. Herzog et al., "The Prediction of Preferences for Familiar Urban Places," *Environment and Behavior* 8 (1976): 627-645.

[24] R. Kaplan, "The Green Experience," in *Humanscape: Environments for People,* ed. S. Kaplan and R. Kaplan (Ann Arbor: Ulrich's Books, 1982), pp. 186-193.

[25] F. Ermuth, *Residential Satisfaction and Urban Environmental Preferences,* York University Department of Geography, Geographical Monograph no. 3 (Toronto: York University Department of Geography, 1974); V.K. Smith, "Residential Location and Environmental Amenities," *Regional Studies* 11 (1977): 47-61; G.R. Hovinen, "Leapfrog Developments in Lancaster County: A Study of Residents' Perceptions and Attitudes," *Professional Geographer* 24 (1977): 194-199; D.I. Patel, *Exurbs: Urban Residential Development in the Countryside* (Washington, D.C.: University Press of America, 1980); P. Weichhart, "Assessment of the Natural Environment: A Determinant of Residential Preferences," *Urban Ecology* 7 (1982/83): 325-342; Fabos et al., *Frederick Law Olmsted, Sr.*

City System Evolution and Metropolitan Development

In outlining the concept of counterurbanization, Berry and Gillard rely on a set of elemental themes of American culture that are largely based on de Crèvecœur's writings of more than two hundred years ago.[26] The themes are love of newness, nearness to nature, freedom to move, individualism, the idea of the melting pot, violence, and a sense of destiny; as such they are easily placed into Zelinsky's compact set of four. The recent trends of declining populations in central cities, continually expanding suburbs, and the once again growing nonmetropolitan areas are considered the reaffirmation of personal and social aspirations and preferences that are basic to American culture. The dramatic rise to prominence of the metropolis was but an interlude temporarily suppressing fundamental attitudes and values embedded in the nation's psyche. Accordingly, as counterurbanization supersedes urbanization, it allows the cultural themes to resurface and assert themselves. By reversing the historic process of population concentration and redistributing growth to smaller places with lower densities and richly endowed natural environments, population deconcentration is the spatial manifestation of these mainstream traits.

Surveys of recent migrants to nonmetropolitan areas have established the relative importance of environmental amenities as pull factors in a household's decision to move.[27] For the most part, environmental amenities remain largely undifferentiated and are generally characterized as factors unrelated to job or income. In some cases, the presence and level of natural amenities are inferred by proxy measures, such as density of motels,

---

[26] B.J.L. Berry and Q. Gillard, *The Changing Shape of Metropolitan America: Commuting Patterns, Urban Fields, and Decentralization Processes* (Cambridge, Mass.: Ballinger, 1977); also see B.J.L. Berry, "The Counterurbanization Process: How General?" in *Human Settlement Systems*, ed. N.M. Hansen (Cambridge, Mass.: Ballinger), ch. 2; J.H. St.John de Crèvecœur, *Letters from an American Farmer* (London: Thomas Davies, 1782).

[27] K.P. Nelson, "Urban Economic and Demographic Change," *Research in Urban Economics* 4 (1984): 25-49; J. Kim, "Factors Affecting Urban-to-Rural Migration," *Growth and Change* 14 (1983): 38-43; A.J. Sofranko and F.C. Fliegel, "Neglected Components of Rural Population Growth," *Growth and Change* 14 (1983): 42-49; A.J. Sofranko, "Urban Migrants to the Rural Midwest: Some Understandings and Misunderstandings," in *Population Redistribution in the Midwest*, ed. C.C. Roseman et al. (Ames: Iowa State University Press North Central Regional Center for Rural Development, 1981); J.D. Williams and D.B. McMillan, "Migration Decision Making among Nonmetropolitan-bound Migrants," in *New Directions in Urban-Rural Migration,* ed. D.L. Brown and J.M. Wardwell (New York: Academic Press, 1980), ch. 8; D.A. Dillman, "Residential Preferences, Quality of Life, and the Population Turnaround," *American Journal of Agricultural Economics* 61 (1979): 960-966; W. Zelinsky, "Coping with the Migration Turnaround: The Theoretical Challenge," *International Regional Science Review* 2 (1977): 175-177; G.V. Fuguitt and J.J. Zuiches, "Residential Preferences and Population Distribution," *Demography* 12 (1975): 491-504.

tourist expenditures, number of secondary homes per capita;[28] in others, town size and population density are used to substitute for direct measurements of environmental amenities.[29]

Where Berry and Gillard see a simple duality in American urban history that is informed by cultural values and ideals, Simmons identifies a total of four different models of urban system development and organization.[30] Each has a distinctive spatial structure and each features characteristic paths of urban growth. The frontier-mercantile and staple-export models are based on the local exploitation and subsequent shipment of raw materials. In the first model urban growth is dictated by an urban core region whose internal hierarchical structure is weakly developed and consists of a few primary cities. In the staple-export model sustained production surpluses provide the lever for skeletal urban development within a region whose overall growth depends on continued external demands for its staple products. Industrial specialization defines a third, essentially aspatial model as lateral and crisscrossing linkages transform previously hierarchical structures. Growth within the system is tied to economies of scale and agglomeration, industrial structure, capital investment, entrepreneurial competence, and the ability to introduce as well as adopt innovations.

Simmons's last model, that of social change, describes a system within which phases of growth are not only dependent on economic conditions, but increasingly on factors of social well-being and on quality-of-life issues. Rising incomes and education, changing preferences for residential locations and new life styles, and environmental amenities are the principal components that account for recent shifts in the distribution of urban populations. The location of population growth as prescribed by this

---

[28] R. Lamb, *Metropolitan Impacts on Rural America*, University of Chicago Department of Geography Research Paper no. 162 (Chicago: University of Chicago Department of Geography, 1975); R.L. Morrill, "Bases for Peripheral Urban Growth," in *Impact of Urbanization and Industrialization on the Landscape*, ed. D.R. Deskins, University of Michigan Department of Geography, Michigan Geographical Publication no. 25 (Ann Arbor: University of Michigan Department of Geography, 1980); D.T. Lichter and G.V. Fuguitt, "The Transition to Nonmetropolitan Population Deconcentration," *Demography* 19 (1982): 211-221.

[29] C.C. Roseman, "Exurban Areas and Exurban Migration," in *The American Metropolitan System*, ed. S.D. Brunn and J.O. Wheeler (New York: Wiley, 1980), ch. 4; T. Heaton et al., "Residential Preferences, Community Satisfaction, and the Intention to Move," *Demography* 16 (1979): 565-573; D.A. Dillman and R.P. Dobash, *Preferences for Community Living and Their Implications for Population Distribution*, Washington State University Agricultural Experiment Station Research Bulletin no. 764 (Spokane: Washington State University Agricultural Experiment Station, 1972).

[30] J.W. Simmons, "The Organization of the Urban System," in *Systems of Cities*, ed. L.S. Bourne and J.W. Simmons (New York: Oxford University Press, 1978), pp. 61-69.

model is highly unpredictable, such that the factors of growth can be recognized only with hindsight. Growth depends on information and technological change. As it responds to the removal of technical and physical constraints, it remains relatively unstable.

The model's developmental sequence focuses on the organizational, functional structure of national urban systems and how that structure defines locations of growth, stagnation, and decline. Bartels distinguishes two additional approaches to city system evolution.[31] One is concerned primarily with the spatial dimensions of a system and looks to detect systematic patterns in the spatial distribution of change. The other examines specific factors, such as resource base and mix, transportation technology, or supply of labor and capital, without necessarily considering their direct or indirect effects on the organizational structure of the system. Borchert's interpretation of metropolitan evolution in America follows the last approach as he reviews the passage of ten innovations in the technology of transport and industrial energy.[32] The last of four developmental periods since 1790 is called the auto-air-amenity epoch. The characteristic pattern of growth is that of regional and metropolitan dispersal. The environmental amenities associated with growth are cursorily identified as warm climates at the regional level, and open space, panoramic views, and clean air at the local level.

Both Simmons's and Borchert's developmental divisions can be separated into two groups that exhibit a temporal dichotomy resembling Berry and Gillard's proposal. In each case the first three stages characterize the development of a national system of cities that continued to push its settlement frontier westward. Both authors agree that since the frontier disappeared as a component of settlement expansion, the system's internal organization and spatial structure have been reworked into a fully integrated national system.[33] Across the nation as well as within individual metro-

---

[31] D. Bartels, "Theorien nationaler Siedlungssysteme und Raumordnungspolitik," *Geographische Zeitschrift* 67 (1979): 110-146.

[32] J.R. Borchert, "American Metropolitan Evolution," *Geographical Review* 57 (1967): 301-332; also see E.S. Dunn Jr., *The Development of the U.S. Urban System*, vol. 1 (Baltimore: Resources for the Future, 1983). Borchert prefaces his discussion by stating: "There is, of course, no implication that the technological changes have been independent variables or basic causes of growth. The presumptions are, rather, that within the given framework of values and institutions they in turn not only further stimulated economic growth but also helped to differentiate it geographically" (p. 302).

[33] A discussion of how to approach the broader issues of change in urban systems is given by M.P. Conzen, "The American Urban System in the Nineteenth Century," in *Geography and the Urban Environment: Progress in Research and Applications,* vol. 4, ed. D.T. Herbert and R.J. Johnston (London: Wiley and Sons, 1981), pp. 316-320.

politan regions the distribution of relative gains and losses is influenced by a distinctly different assortment of variables. Among them are education and research, government investment at the federal and state levels, and environmental amenities; however, there appears to be no coherent, locational logic that binds them, if it were not for the inherent inertia of accumulated structures.[34]

Borchert's discourse may lack in specificity, but it provides the only descriptive bridge that links studies of interurban relationships and structures to those concerned with intraurban patterns and growth processes.[35] Adams classifies the pattern of residential growth in urban areas according to four significantly different transport eras.[36] In sequence each era represents a critical improvement of intraurban movement and accessibility. Two of these, the walking/horse car era and the recreational auto era, generate movement surfaces that are associated with a more compact, nearly circular urban form. The other two, the electric streetcar era and the freeway era, channel movement within a well-defined linear network extending urban growth along narrow bands into the surrounding rural area. Adams's schematic presentation of alternating periods of transportation-induced metropolitan growth serves as a corrective feature of Clark's population density model. As far as the Midwestern city is concerned, the intraurban pattern of concentric rings, representing exponentially declining population densities away from the city center, are compromised by an axial pattern of distended peripheral growth following commuter rail lines and urban freeways.

Adams's conception of the historic sequence of urban growth does not deal with environmental amenities as significant factors of urban expansion. In fact, the impact of recreational automobile use was such as to consolidate the urban form to near concentricity. Nor does it offer a perspective that invites interregional comparisons. Granted, its similarity with Borchert's argument may be superficial, yet its emphasis on accessibility and

---

[34] J.R. Borchert, "America's Changing Metropolitan Regions," *Annals of the Association of American Geographers* 62 (1972): 352-373.

[35] Borchert clearly recognizes the need to establish a conceptual bridge between the two scales. Bartels stresses the importance of such linkage particularly in light of continued suburban growth and expansion, which has begun to erase the conventional distinctions of local and regional scales to create the "galactic metropolis." Unfortunately, the theoretical or conceptual base of such linkage does not yet exist. See Bartels, "Theorien nationaler Siedlungssysteme," p. 111; P.F. Lewis, "The Galactic Metropolis," in *Beyond the Urban Fringe,* ed. R.H. Platt and G. Macinko (Minneapolis: University of Minnesota Press, 1983), pp. 24-49.

[36] J.S. Adams, "Residential Structure of Midwestern Cities," *Annals of the Association of American Geographers* 60 (1970): 37-62.

varying transportation surfaces within urban areas brings it in direct contact with economic theory. If then the national settlement patterns are shaped by the presence of, and demand for, natural resources, why should not the form of urban areas, expressed in terms of population density and intensity of land use, be influenced by the distribution of locational attributes other than distance? If the national system is presently reshaped by a massive yet selective redistribution of people, are similar events taking place at the local, metropolitan level? If economic considerations are less important in the accounting of national migration, why should they still matter at the local level, such as in the form of a simple trade-off between savings on transportation costs and residential space?

Paradoxically, even as urban structure is referenced to rational economic behavior, economic motivations are considered largely irrelevant when it comes to intraurban mobility and residential preferences.[37] During the last twenty years economists have taken steps to incorporate additional site attributes, aside from distance to a city's center, into their analysis of property values and patterns of residential relocation.[38] The theoretical link between standard bid-rent models and the urban population density model are well documented.[39] To explore the effects of environmental amenities on the distribution of population within urban areas appears to be an obvious extension. The specific procedures and operational definitions of the environmental amenities considered in this study are discussed in the following chapter.

---

[37] J.W. Simmons, "Changing Residence in the City: A Review of Intraurban Mobility," *Geographical Review* 58 (1968): 622-651; J.M. Quigley and D.H. Weinberg, "Intra-Urban Residential Mobility: A Review and Synthesis," *International Regional Science Review* 2 (1977): 41-66; S.E. White, "The Influence of Urban Residential Preference on Spatial Behavior," *Geographical Review* 71 (1981): 176-187; W.A.V. Clark, "Recent Research on Migration and Mobility," *Progress in Planning* 18 (1982): 1-56; W.A.V. Clark and J.L. Onaka, "Life Cycle and Housing Adjustment as Explanations of Residential Mobility," *Urban Studies* 20 (1983): 47-57.

[38] P. Linneman, "Hedonic Prices and Residential Location," in *Economics of Urban Amenities*, ed. D.B. Diamond and G.S. Tolley (New York: Academic Press, 1982), ch. 3; Q. Gillard, "The Effect of Environmental Amenities on Home Values: The Example of the View Lot," *Professional Geographer* 33 (1981): 216-220. A more general discussion of how to incorporate environmental goods and services not traded on the market is provided by P.-O. Johansson, *The Economic Theory and Measurement of Environmental Benefits* (Cambridge: Cambridge University Press, 1987).

[39] I. Orishimo, *Urbanization and Environmental Quality* (Boston: Kluwer Nijhoff, 1982); H.W. Richardson, *Urban Economics* (Hinsdale: Dryden Press, 1978).

Chapter 4

METHODOLOGY, DATA BASE, AND OPERATIONAL DEFINITIONS

> *Eine Wissenschaft der Entfernungen ist eines der ersten Erfordernisse der Geographie als Wissenschaft der räumlichen Anordnungen auf der Erdoberfläche.*
> —F. Ratzel, *Erdenmacht und Völkerschicksal*

> [T]he "fetishism of space" has been realised: relatively few claims are made now that geography is a "discipline in distance."
> —R.J. Johnston, "Ideology and Quantitative Human Geography"

Method of Analysis

Stepwise multiple-regression procedures are used to identify and estimate the effects of natural landscape features on suburban population densities as well as to determine their relative importance vis-à-vis three measures of intrametropolitan access. The underlying analytical model belongs to the larger family of urban population density functions. Population density has been the most commonly and readily analyzed aspect of urban form. Cartographic representations and simple cross-sectional profiles have long provided the conventional means to illustrate and analyze the generally observed phenomenon of urban population densities diminishing with distance from the city center. As an alternative model, the negative exponential function, tested in a wide range of cases, has been found to approximate competently the spatial variation of density in cities.[1]

---

[1] R.E. Groop and J.-C. Muller, "Evaluating an Urban Model: A Cartographic Approach," *American Cartographer* 5 (1978): 111-120. The authors use standard one-dimensional cartographic techniques to show the existence of exponential density gradients.

In the strictest sense, this study adheres to the model's basic premise. However, by adding a number of directional and accessibility qualifications as well as a series of landscape features and conditions, it alters substantially the function's simple, two-parameter specification. In principle, this study softens, if not severs, the unilateral link between density and distance. Instead of expecting the logarithm of population density to decline at a constant rate, the study's primary hypothesis allows for irregular rates of decline and localized reversals. The extent of gradient variation is assumed, at least in part, to be a function of desirable locational amenities. A number of surficial natural landscape features, previously identified as potentially significant amenities, are tested to estimate their effect on suburban population densities.

Origin and Use of Density Functions

The study of population density gradients has been a familiar element in the analysis of urban form. Common to most attempts at understanding how cities grow and expand is the notion that urban population densities are inversely related to distance with regard to a given city center according to the equation

$$D_d = D_0 e^{-bd} \qquad (1)$$

where D_d is the density at distance d from the metropolitan center, D_0 is the density at the center, and b is the density gradient or the rate at which density diminishes with distance; e is the base of the natural logarithm. The negative exponential function has been the most frequently used and tested specification of the empirical relationship between density and distance.[2] It provides robust interval-level measurements that characterize not only the nature of population distribution within urban areas, but also changes over time. For example, simultaneous decreases in gradient and central density have been used to define the rate and extent of suburbanization (deconcentration) and overall compactness (congestion) of urban areas.[3]

When first introduced, the negative exponential function was unencumbered by any rigorous theoretical arguments concerning urban spatial structure. It opened a convenient shortcut toward efficiently summarizing

---

[2] B. Edmonston, *Population Distribution in American Cities* (Lexington, Mass.: Lexington Books, 1975); J.F. MacDonald and H.W. Bowman, "Some Tests of Alternative Urban Population Density Functions," *Journal of Urban Economics* 3 (1976): 242-252.

[3] H.H. Winsborough, "City Growth and City Structure," *Journal of Regional Science* 4 (1962): 35-49; H.H. Winsborough, "An Ecological Approach to the Theory of Suburbanization," *American Journal of Sociology* 63 (1963): 565-570; A.M. Guest, "Population Suburbanization in American Metropolitan Areas, 1940-1970," *Geographical Analysis* 7 (1975): 267-283; M.A. Goldberg and J. Mercer, *The Myth of the North American City* (Vancouver: University of British Columbia Press, 1986), pp. 151-152.

observed urban density patterns.[4] As a descriptive tool it facilitated cross-sectional, cross-cultural, and longitudinal comparisons. Furthermore, it provided a diagnostic scale that pegged long-term trends of urban growth and development. The specific form of density gradient as negative exponential received early critical support from microeconomic studies that offered empirical evidence and theoretical support concerning the essential similarity between urban rent gradients and population density gradients.[5]

From this juxtaposition emerged a common set of presumptive conditions that compress urban form and structure into a static, one-dimensional model: located on a featureless plain, cities are, for the sake of argument, considered monocentric, circular, and compact. Central employment allows travel to consist solely of commuting to work. Population is given exogenously. Since urban residents share the same utility and demand functions, the size of cities is principally determined by incomes, tastes, and residential preferences. In short, the exponential density decay reflects a spatial series of equilibrium conditions controlled by the direct trade-off between land (rent and residential space) and accessibility (commuting cost). The resulting pattern of residential location allows both consumers and producers to maximize their respective utility and profits.[6]

After a period of extensive testing, duplication, and verification, the study of population density functions and gradients has followed two separate paths. The first sought to develop alternative density functions better fitting the observed pattern of urban population densities. While retaining distance as the sole independent variable, researchers proposed and experimented with numerous specifications of density and distance.[7] The second-degree polynomial (quadratic exponential) was the first attempt to reconcile

---

[4] Convenient indeed, for even some of the critics who reject the model's underlying contextual reductionism employ gradient analysis to support a series of sweeping generalizations concerning a comparison of Canadian and U.S. American cultural traits. Does cultural uniformity exist if there is no substantial difference in gradient to be found? See B. Edmonston, M.A. Goldberg, and J. Mercer, "Urban Form in Canada and the United States: An Examination of Urban Density Gradients," *Urban Studies* 22 (1985): 209-217; Goldberg and Mercer, *The Myth of the North American City*, pp. 151-152; C. Clark, "Urban Population Densities," *Journal of the Royal Statistical Society* 114 (1951): 490-496.

[5] B.J.L. Berry, "Cities as Systems within Systems of Cities," in *Regional Development and Planning*, ed. J. Friedman and W. Alonso (Cambridge: Massachusetts Institute of Technology Press, 1964), ch. 6; I. Orishimo, *Urbanization and Environmental Quality* (Boston: Kluwer Nijhoff, 1982), ch. 3.

[6] R.F. Muth, *Cities and Housing* (Chicago: University of Chicago Press, 1969); E.S. Mills, *Studies in the Structure of the Urban Economy* (Baltimore: Johns Hopkins University, Resources for the Future, 1972).

[7] For a concise overview see MacDonald and Bowman, "Some Tests."

the crater of central densities.[8] More recent studies have employed switching regressions and cubic spline functions in order to accommodate apparent discontinuities in the spatial pattern of metropolitan densities.[9] Others have argued that the emergence of polycentric metropolitan areas effectively will override the general usefulness of the simple density gradient.[10] However, determining the location of subcenters, "knobs," "switches," or "turning-points" has been a persistent problem in developing convincing and comprehensive alternatives.

Research along the second path set out to build a more coherent foundation consistent with standard (neoclassical) land-use theory. Rigorous economic arguments were presented either to justify or reject particular density functions. Thus, the much-used negative exponential form obtained only under severe restrictions concerning the nature of commuting costs, consumer tastes, and housing production technology. Other factors, marketable or not, were ignored, or kept constant, even though their general importance in the metropolitan housing market was recognized. In general, the tendency to hold on to a few, oft-tested expressions of the density function, in particular the exponential, remained strong.[11] "Distance" was a powerful spatial catchall term that could absorb and reflect major elements of the economic structure of cities. Experimentation with various operational definitions of density either removed or clarified some of the more restrictive conditions.[12] Derivative density and distance measures were substituted for unavailable data or mimicked inferred externalities. However by adding transformed distance

---

[8] B. Newling, "The Spatial Variation of Urban Population Densities," *Geographical Review* 59 (1969): 242-253; R.F. Latham and M.H. Yeates, "Population Density Growth in Metropolitan Toronto," *Geographical Analysis* 2 (1970): 177-185.

[9] J.K. Brueckner, "A Switching Regression Analysis of Urban Population Densities: Preliminary Results," *Papers of the Regional Science Association* 56 (1985): 71-87; J.E. Anderson, "The Changing Structure of a City: Temporal Changes in Cubic Spline Urban Density Patterns," *Journal of Regional Science* 25 (1985): 413-425; A. Okabe and S. Masuda, "Qualitative Analysis of Two-Dimensional Urban Population Distributions in Japan," *Geographical Analysis* 16 (1984): 301-312.

[10] D.A. Griffith, "Modelling Urban Population Density in a Multi-centered City," *Journal of Urban Economics* 9 (1981a): 298-310; D.A. Griffith, "Evaluating the Transformation from a Monocentric to a Polycentric City," *Professional Geographer* 33 (1981): 189-196; R.A. Dubin and C.-H. Sung, "Spatial Variation in the Price of Housing: Rent Gradients in Nonmonocentric Cities," *Urban Studies* 24 (1987): 193-204.

[11] J.K. Brueckner, "A Note on Sufficient Conditions for Negative Exponential Population Densities," *Journal of Regional Science* 22 (1982): 353-359; MacDonald and Bowman, "Some Tests."

[12] H.W. Richardson, "On the Possibility of Positive Rent Gradients," *Journal of Urban Economics* 4 (1977): 246-257; R.P. Numrich, *Essays on the Econometric Estimation of Urban Spatial Structure* (Ph.D. diss., State University of New York at Albany, 1986).

and density terms, the identity of locational amenities, which presumably were to counteract the paramount negative effect of distance, remained dubious.[13] So, despite the ongoing debate over which functional specification is most appropriate, the basic distance-density relationship has been left intact.

Instead of emphasizing gradient analysis in a comparative framework of urban development or searching for the theoretically optimal functional form of urban density decay, some researchers have identified and isolated specific environmental factors and conditions that, as amenities, may further articulate, or even cancel, the dominant influence of distance to city center. Distance expressed in terms of accessibility is joined by a growing list of additional locational amenities that influence the distribution of urban population and urban land values.[14] No longer mandating the theoretical existence of a universal urban structure, recent interpretations of urban form and growth have taken specific notice of local variations in amenity endowment. Research of this kind has been argued mostly within the linear framework of multiple regression; as such the inferred effects of amenities could be attached to the exponential terms of Clark's density equation.

Estimation Form of the Modified Density Function

The logarithmic transformation of Clark's negative exponential function yields the following equation

$$ln\ D_d = ln\ D_0 - bd. \qquad (2)$$

Its parameters can be estimated by ordinary least squares. The modified multiple-regression model used in this study follows in principle the example of previous research concerned with the effects of locational amenities on the distribution of urban populations.[15] The model's generalized equation, substituting c (= constant or intercept) for $ln\ D_0$, is:

---

[13] A. Rodriguez-Bachiller, "Discontiguous Urban Growth and the New Urban Economics: A Review," *Urban Studies* 23 (1986): 79-104.

[14] G. Alperovich, "Neighborhood Amenities and Their Impact on Density Gradients," *Annals of Regional Science* 14.2 (1980): 51-64; D.B. Diamond, "The Relationship between Amenities and Urban Land Prices," *Land Economics* 56 (1980): 21-32; R. Pollard, "View Amenities, Building Heights, and Housing Supply," in *The Economics of Urban Amenities*, ed. D.B. Diamond and G.S. Tolley (New York: Academic Press, 1982), ch. 5; Q. Gillard, "The Effect of Environmental Amenities on Home Values: The Example of the View Lot," *Professional Geographer* 33 (1981): 216-220; J.P. Briscoe, "Statistical Relationships of Growth in Urban Population Densities as Related to Selected Controllable Environmental Factors" (Ph.D. diss., University of Louisville, 1977).

[15] R.L. Morrill, "Bases for Peripheral Urban Growth," in *Impact of Urbanization and Industrialization on the Landscape*, ed. D.R. Deskins, University of Michigan Department of Geography, Michigan Geographical Publication no. 25 (Ann Arbor: University of Michigan

$$ln\ D_d = C + S_i b_i x_i \text{ for } i = 1, 2, ..., n \tag{3}$$

where n is the number of regressors that in addition to a single linear distance term include several access and landscape variables. Except for distance, the specific direction and relative magnitude of all additional variables are at this stage fairly indeterminate, although this study's hypothesis proposes that physiographic features and conditions have a positive impact on densities. Of the two qualitative accessibility terms, suburban commuter railroad access is expected to have had a cumulative and positive effect on population densities. The spatial role of expressways, on the other hand, is less clearly defined in spatial models of urban growth. The prevalent use of the private car by suburban commuters is credited with contributing to the far-flung, even diffuse residential patterns of suburban growth and expansion.[16] Stepwise regression is used to calibrate each particular equation. The order of variable entry is similar to the forward-selection technique in that the stepwise method begins with a bivariate equation and accepts additional variables one at a time; each entry depends on the variable's relative importance and contribution to the model. At each step all independent variables, including those already entered, are once more considered, allowing for the removal of previously accepted regressors.

Inasmuch as this method of analysis complements the research strategies used in the study of urban housing and land prices, there are other reasons why additional parameters of linear distance are not included. Since the study area is limited to the suburban portion of the Chicago metropolitan area, there is no need to add a quadratic distance term in order to account for a central density crater and locate the crest of the density gradient. Essentially exponential conditions obtain on the suburban side of the crest; that is, the natural logarithm of population density falls at a fairly constant rate.[17] Furthermore, to enter several transformed distance terms (e.g., quadratic and cubed) may ultimately obscure or subsume the effect of yet unspecified locational amenities or disamenities.[18]

A solution to this dilemma is to specify from the outset the order in which the variables enter the equation. However, the literature dealing with the identification and evaluation of natural landscape features and conditions has produced no consensus as to the relative order and signifi-

---

Department of Geography, 1980), pp. 95-117; Alperovich, "Neighborhood Amenities"; Diamond, "The Relationship between Amenities."

[16] Goldberg and Mercer, *The Myth of the North American City*, pp. 152-154.

[17] J.B. Parr, "A Population-Density Approach to Regional Spatial Structure," *Urban Studies* 22 (1985): 289-303.

[18] R.F. Lamb, *Metropolitan Impacts on Rural America*, University of Chicago Department of Geography Research Paper no. 162 (Chicago: University of Chicago Department of Geography, 1975).

cance of individual landscape components, let alone their effect on the distribution of urban population. Because of the general lack of a theoretical rationale, each natural landscape variable is considered a potential amenity. Finally, possible intraregional variation in the effect of radial distance is allowed for by two access variables and a directional qualification that divides the suburban region into several sectors. Such spatial qualifiers remove the need for additional distance terms and favor a single distance term in its undiluted form.

Selection of Variables and Data Sources

The variables considered for this study came under three major headings: density (population/land area), accessibility, and physiography. The variables needed to be compatible in terms of both time and area covered. Comprehensive coverage of the suburban area of a metropolitan region was also desirable. Most important, the data sets needed to be internally consistent, which allowed some flexibility of spatial aggregation and separation without disrupting the essential locational matches among all variables.

This investigation focuses on the visible, surficial, fairly permanent components of the natural landscape. The term "natural" is used in order to exclude the brick-and-mortar components of the cultural landscape. No attempt was made to distinguish between indigenous and planted vegetation or between lake and impoundment. The natural landscape variables were selected according to their potential for circumscribing the general character of a "rural Eden," which is defined as a parklike blending of landscape elements, neither wilderness nor city.[19] More specifically, the variables should be able to capture an environmental quality that presents a certain sense of orderliness, yet recognizes in sufficient detail significant landscape shapes, land cover, and land uses, with variation in terrain, vegetation, surface water, and land suited for outdoor recreational activities being the most important attributes. These landscape components, together or separately, have repeatedly been identified as factors that most consistently affect a person's environmental preferences. They also appear to be part of a set of cultural predispositions and social aspirations that have given this nation its distinctive suburban history and perspective.

Straight-line distance from the metropolitan center, commuter railroads as public transportation, and expressways for private automobile use

---

[19] W. Zelinsky, *The Cultural Geography of the United States* (Englewood Cliffs: Prentice Hall, 1973), p. 64. For a more detailed discussion and literature review see chapter 2. The problems and issues of landscape evaluation research are presented by several authors in a special edition of the *Transactions of the Institute of British Geographers* (pp. 119-161). The collection is introduced by J. Appleton, "Landscape Evaluation: The Theoretical Vacuum," *Transactions of the Institute of British Geographers* 66 (1975): 120-123.

are the three major components of intrametropolitan accessibility. The importance of the last two is routinely acknowledged as falling density gradients are promptly associated with improvements in transportation technology and attendant reductions in costs.[20] Population counts and settled land area are crucial elements for estimating the negative exponential density function. The methods of acquiring the data for both variables range from use of existing census tract enumerations to the two-point estimation procedure that uses only two population totals.[21] However, to preserve both radial and directional variations in density, the concentric zones must be of equal width and the accounting cells within the zones, of equal size.[22]

Many of these conditions and requirements are at least partially fulfilled by a series of data files that have been compiled, edited, and maintained by the Northeastern Illinois Planning Commission (NIPC). Each file has a spatial resolution of nominally 160 acres, equal to the size of a quarter section of the ubiquitous rectangular U.S. Public Land Survey.[23] The geographic coverage is limited to the suburban portion of the six-county metropolitan planning region centered on Chicago. The Environmentally Sensitive Features (ESF) file contains data on physiographic features that were recorded during the 1970s.[24] As some of the environmental information is based on primary sources (e.g., satellite images, aerial photographs, and field surveys) and some on secondary sources (e.g., topographic maps), differences in scale and interpretation introduced some unavoidable inconsistencies in environmental detail. However, the planimetric measurements or ratings of each environmental variable remain internally consistent. The ESF file uses either areal measurements (rounded to the nearest n x 10 acre) and either ordinal or binary classifications.

Data on population counts, land excluded from settled areas, and total acreage of quarter sections are enumerated in the population and land-

---

[20] For example see E.S. Mills, "Urban Density Functions," *Urban Studies* 7 (1970): 5-20.

[21] J.A. Kahimbaara, "The Population Density Gradient and the Spatial Structure of a Third World City: Nairobi, a Case Study," *Urban Studies* 23 (1986): 314-315; Edmonston, *Population Distribution in American Cities*; Edmonston, Goldberg, and Mercer, " Urban Form in Canada and the United States," pp. 211-213; Mills, "Urban Density," p. 8.

[22] Groop and Muller, "Evaluating an Urban Model," p. 115.

[23] The lasting significance of the rectangular land survey to the structure of suburban development in this country is undeniable. Employing aggregates of sections makes both historical and functional sense. For an excellent discussion on this subject see H.B. Johnson, *Order upon the Land: The U.S. Rectangular Land Survey and the Upper Mississippi Country* (New York: Oxford University Press, 1976).

[24] The sources, definitions, and typologies of the selected features were compiled by editors P. Rasmussen and C. Giagnorio, *Environmentally Sensitive Features File Code Book and Definitions of Variables,* Northeastern Illinois Planning Commission Project no. 0501.01 (Chicago: Northeastern Illinois Planning Commission, 1979).

use files; both files are available for 1970 and 1980. Calculations of relative differences in elevation were based on sets of topographic maps, both 7.5- and 15-minute series with a contour interval of no greater than ten feet. The determination of commuter rail and highway access made use of NIPC's 1:126,720 regional map, first issued in 1974 and revised in 1982.

Spatial Matrix of the Study Area

The original spatial resolution of the quarter section as the smallest areal unit results in a detailed inventory that can provide valuable baseline information to county and municipal agencies. However, in order to present sectoral analyses and a regional overview of the effects of environmental features on suburban population densities, the quarter-section division was nominally replaced by a larger spatial unit, which encompassed four sections or 4 x 4 quarter sections. Systematic aggregation assigned each of the thirty-six sections to a fixed spatial unit so that, for example, only sections 1, 2, 11, and 12 or 3, 4, 9, and 10 were combined, but never 2, 3, 10, and 11. Thus each township contained nine such spatial units (Appendix A). There are several reasons why such aggregation is useful and even necessary.

Aggregation decreases the relative differences in size of the enumeration unit. The spatial alignment of the rectangular land survey in the region is characterized by several irregularities that produce considerable variation in actual quarter-section acreage. The sectional grid of T.41N. in northern Kane and Cook counties is incomplete. Owing to surveying error, the first row of quarter sections and sections gradually disappears so as not to disrupt previously surveyed and recorded townships to the north (figure 2). Truncated quarter sections can be found also along the Wisconsin and Indiana state boundaries and Chicago's city limit. In the southwestern part of the region additional discrepancies in size were created by one of the two Indian treaty boundaries that pass obliquely across the rectangular grid. Along this boundary the sectional lines of two land surveys, conducted at different times, meet and are offset by a measurable margin. Further variation in size occurs to the north along the shore of Lake Michigan.

Spatial aggregation helps diminish inconsistencies in scoring and measurement. It reduces, if not entirely eliminates, the potential of spatially mismatching and separating the variables. It also helps in one regard that is particularly crucial to the method of analysis used in this study. In the case of the quarter-section division it may happen that a given landscape feature, such as a park or a lake, fills the entire area, therefore preempting all land suitable or usable for residential development. Even though population densities in the surrounding quarter section may be high, the spatial link between density and lake would be severed and could not be recognized by the statistical techniques employed here. Since landscape features are

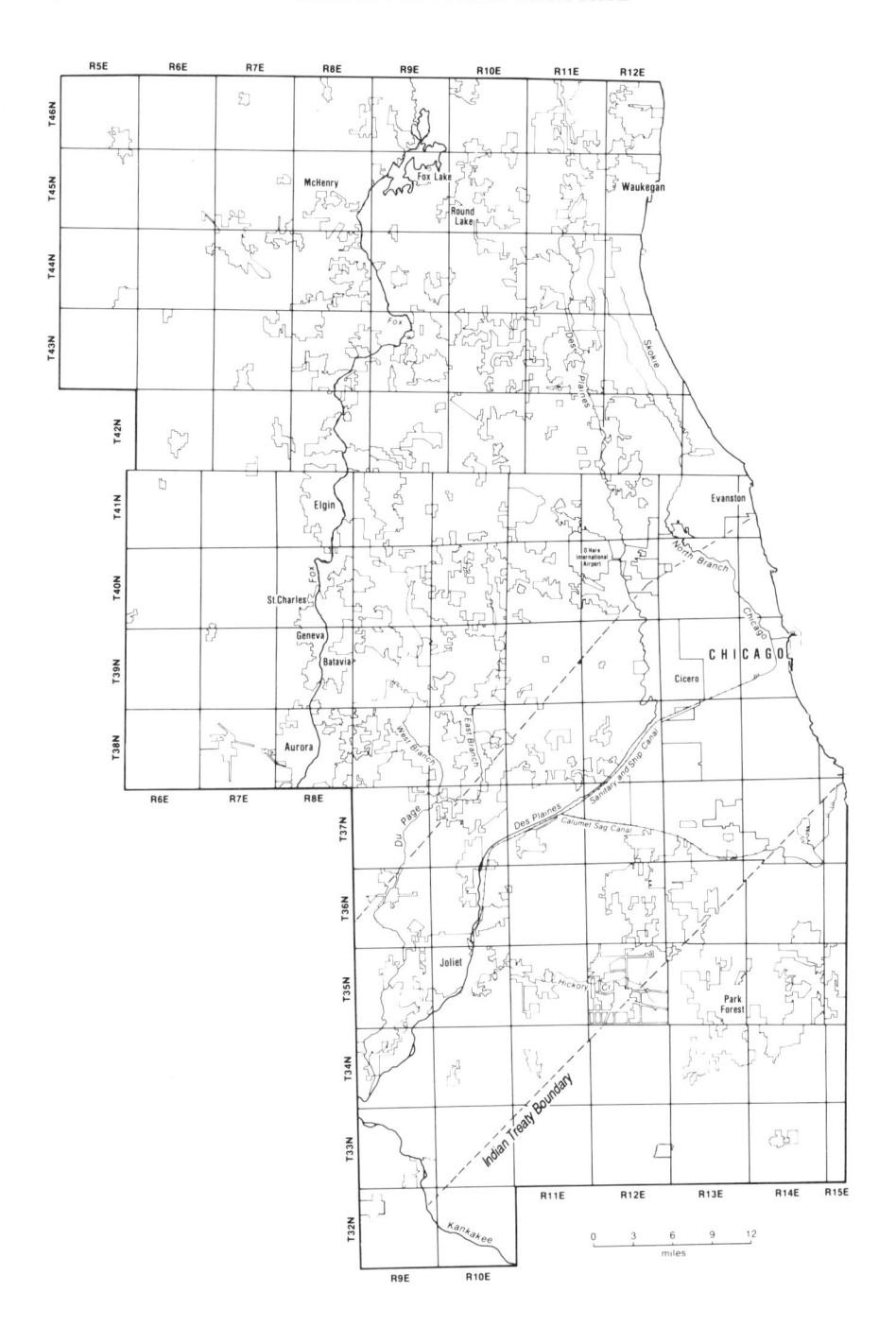

Fig. 2. The geometry of the rectangular land survey in northeastern Illinois.

considered externalities that as locational attributes are independent of population, they should not enter, directly or indirectly, into the calculation of population densities. Thus, aggregation reduces the risk of arbitrarily separating population density from a given environmental feature. Finally, this method of spatial grouping by proximity is an accepted procedure in examining patterns of spatial relationship among several variables.[25] In this case it has the added advantage of increasing computational convenience and decreasing associated costs.

The Variables and Their Operational Definitions

The following list of variables is accompanied by a series of maps that show the distribution of the selected natural landscape variables in the Chicago suburban region. Also shown is the extent of commuter railroad access and highway corridors. The regional maps of 1970 and 1980 population densities can be found in chapter 5. This separation is intentional, since none of the maps were prepared for the explicit purpose of cartographically analyzing the patterns of spatial association between the variables.[26] To facilitate subsequent references, each map bears the boundaries that divide the region into five major sectors and the north-shore margin. Held against the background of existing models of urban growth and development in the American Midwest, the maps illustrate the advantages of the analytical framework used in this study, since it facilitates cross-sectional and longitudinal comparisons of density variability within the metropolitan region. It should be noted that the independent variables remain unchanged for both 1970 and 1980. Environmental features are the relatively more stable, nearly constant components of the suburban landscape, whereas population densities can undergo fairly rapid and sizable shifts. Since the environmental inventory was completed during the 1970s, it is reasonable to use the data for both years.

Independent Variables: Natural Landscape Components

Forest (ESF). Landsat-based maps using 1974-75 satellite imagery provided the sole source for determining location and extent of crown cover of deciduous and coniferous trees and shrubs. Areal measurements were rounded to the nearest n x 10 acres (figure 3).

---

[25] P.J. Taylor, *Quantitative Methods in Geography* (London: Houghton and Mifflin, 1977), ch. 5.

[26] Such comparison is invited by J.F. Hart, "Population Change in the Upper Great Lakes,"' *Annals of the Association of American Geographers* 74 (1984): 221-243. For a criticism of Hart's nonpositivist approach see C.C. Roseman, "Cartographic Analysis of Population Change," *Annals of the Association of American Geographers* 75 (1985): 133-135.

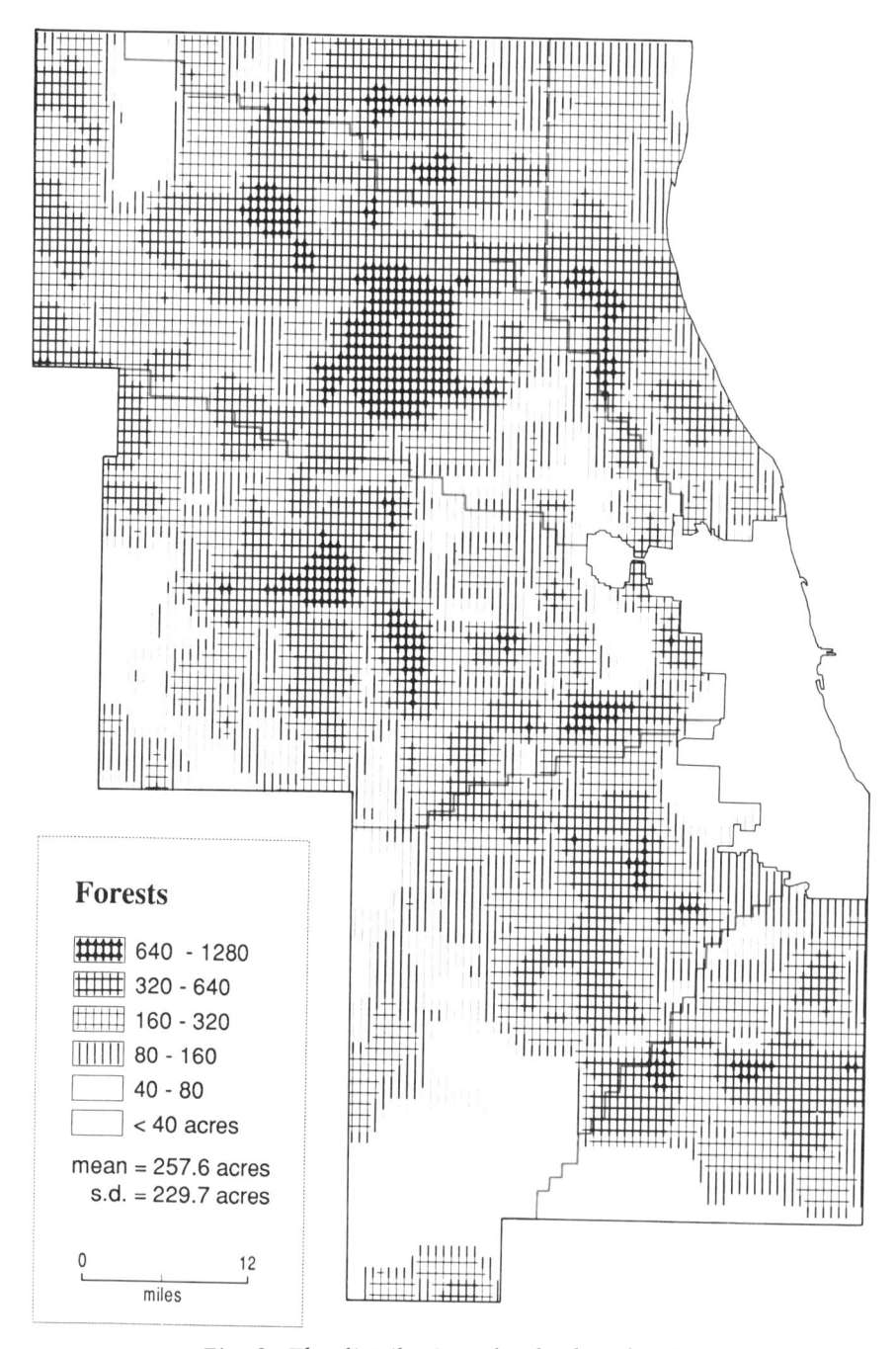

Fig. 3. The distribution of suburban forests.

Nature Park (ESF). Based on 1973 aerial photographs (1:16,000), this variable combines three types of separately recorded nature and park areas. They are conservation areas, nature preserves, and state parks. Planimetric measurements were recorded in multiples of ten acres (figure 4).

Golf Course (ESF). The locations and acreage of golf courses were determined in the same fashion as those of nature parks (figure 5). Unlike the other landscape components, this variable can undergo rapid expansion. In contrast to nature parks or even lakes, golf courses can be constructed in a short period. Extending 1973 data on golf courses to both end points of the decade may seem to dilute the potential of this variable. However, golf courses are now often built at the early stages of residential development and serve as added amenity to attract new residents. Thus, the effect will be felt subsequently, not concurrently.

Lake (ESF). Based on 1974 photo-revised 7.5-minute topographic maps, this variable includes lakes, ponds, and all other artificial impoundments. The minimum size is ten acres. All subsequent measurements were recorded at ten-acre intervals (figure 6).

River (ESF). Based on the same topographic series, rivers were inventoried according to a simple ordinal ranking that classified mapped perennial streams as major and minor. The original scores were changed so that each quarter section containing at least one major stream received a score of 10, the presence of at least one minor stream scored a single point. A total of 160 indicates that a major stream runs through each quarter section of the four-section aggregate (figure 7).

Relief. This variable uses an interval scale of measurement and represents the difference between the highest and lowest elevations within a four-section aggregate. Relief measurements do not represent an average of four individual sections or even quarter sections. The data were taken from either 7.5- or 15-minute topographic maps (depending on availability and contour interval). Since only a few of the 1:24,000 maps have a five-foot interval, the ten-foot contour interval was accepted as the standard. It is common to all 1:62,500 maps for this area and appears on most of the 7.5-minute sheets (figures 8-10).

Instead of converting the cumulative total of planimetric measurements to percentages of total area (so as to account for the still remaining variation in aggregate size), the simple totals of acreage were used in the analysis. Relative measurements were rejected even though initial tests resulted in higher coefficients. With this approach a small lake in a small area has the same value as a large lake in a proportionately larger area. In terms of human perception a small lake stays small regardless of the size of the quarter section. Similarly, it is of little consequence to determine the size of a golf course relative to the size of a four-section aggregate. On the other

Nature Parks and Preservation Areas

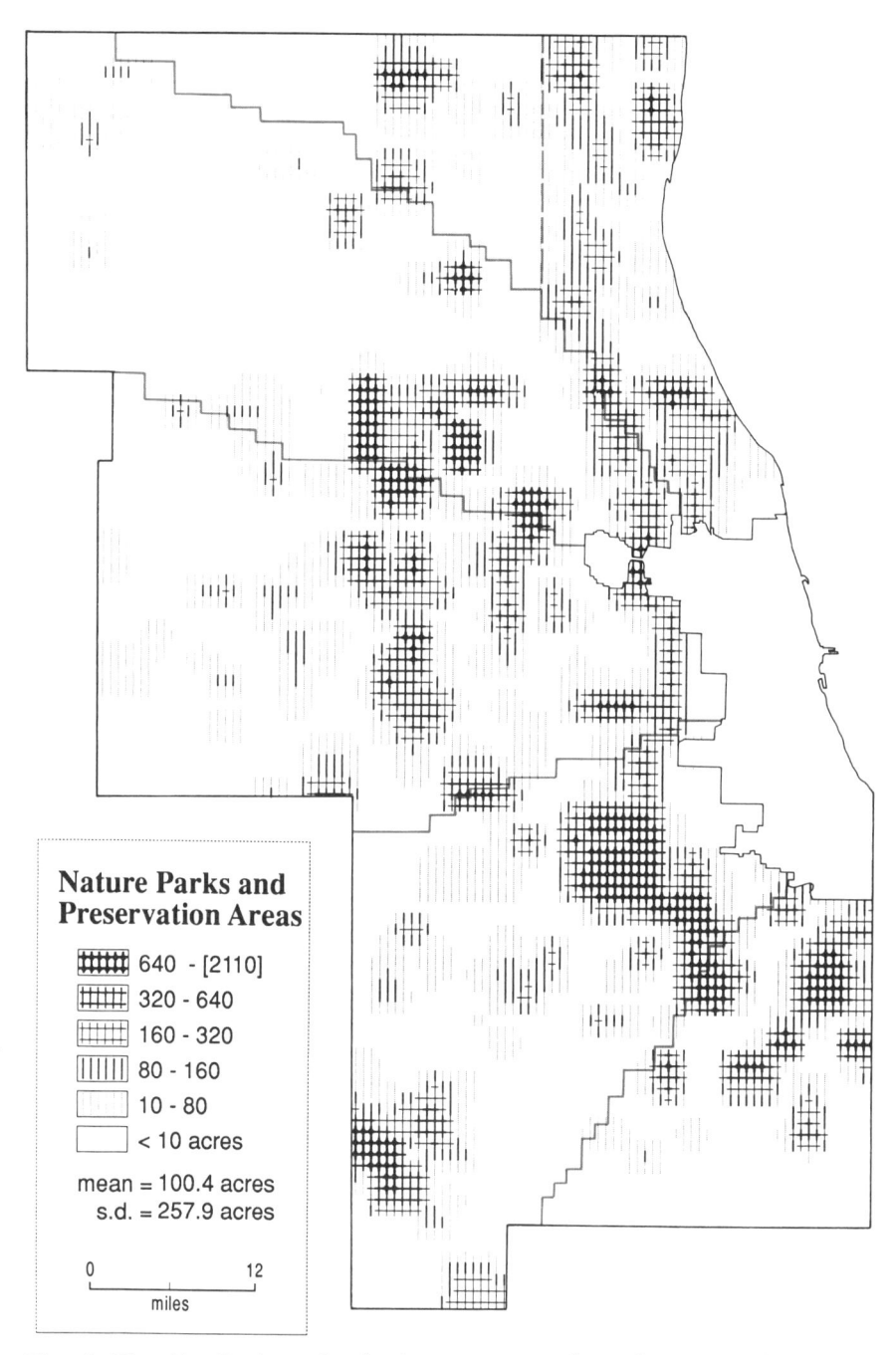

- 640 - [2110]
- 320 - 640
- 160 - 320
- 80 - 160
- 10 - 80
- < 10 acres

mean = 100.4 acres
s.d. = 257.9 acres

0 — 12 miles

Fig. 4. The distribution of suburban nature parks and preservation areas.

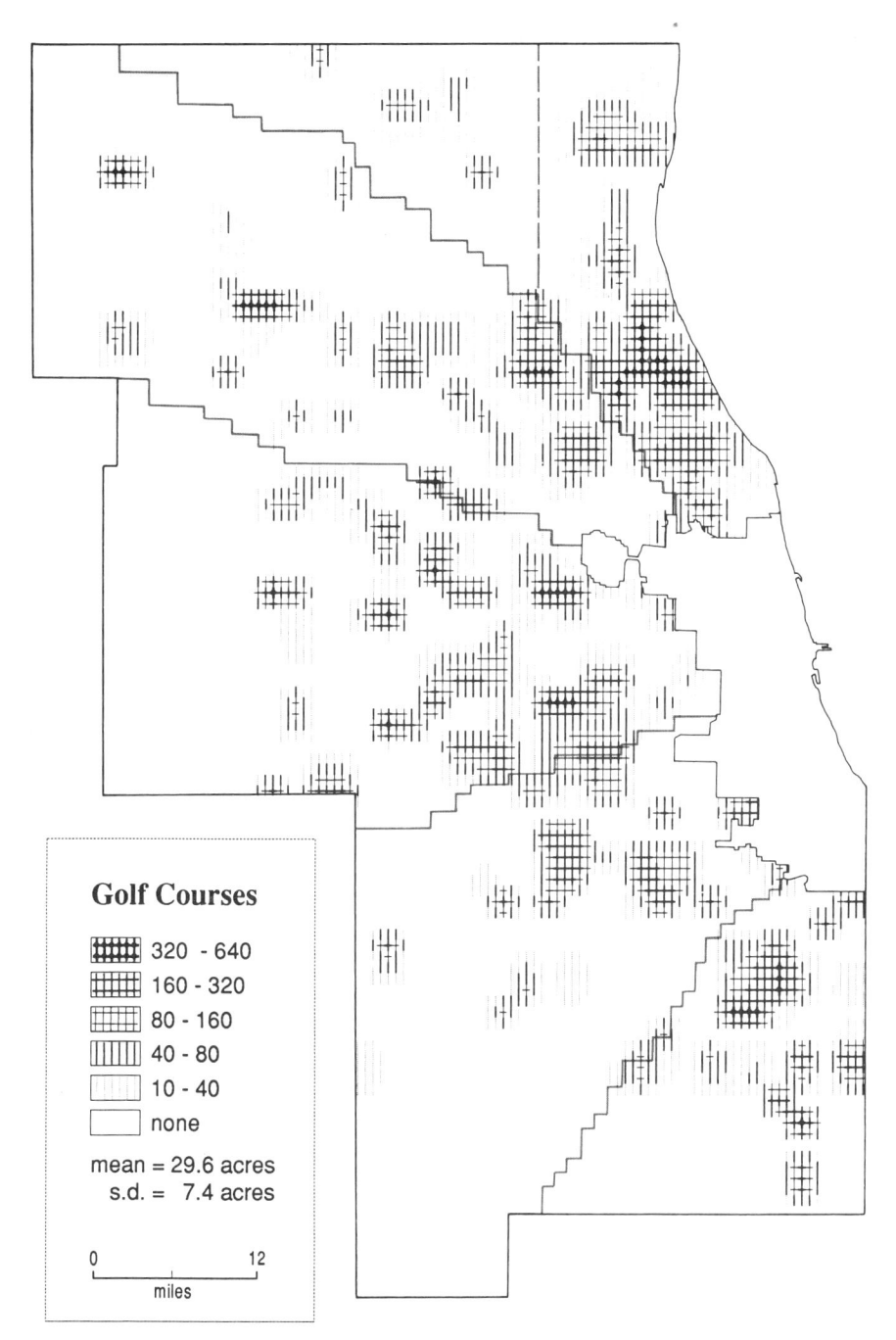

Fig. 5. *The distribution of suburban golf courses.*

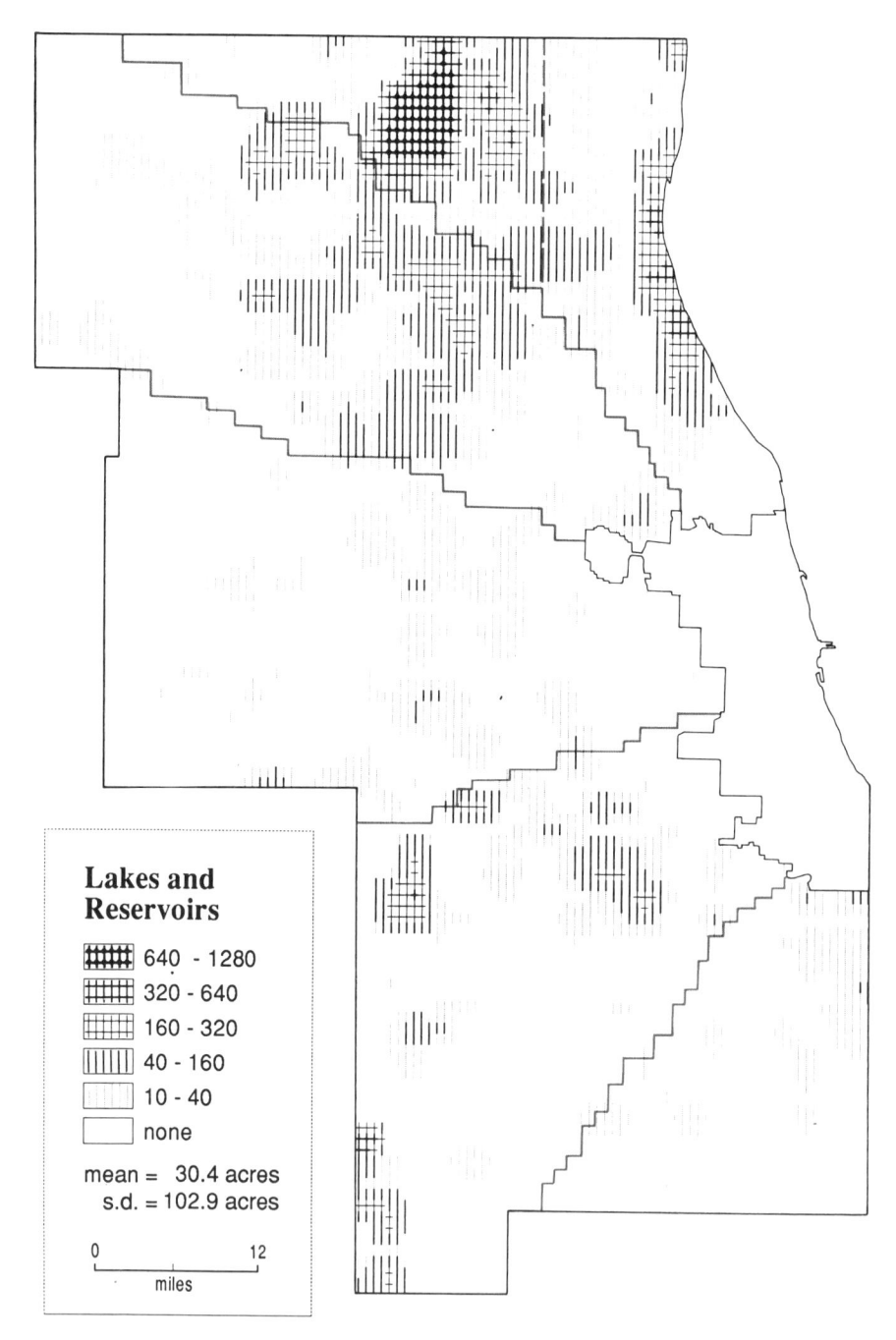

Fig. 6. The distribution of suburban lakes and reservoirs.

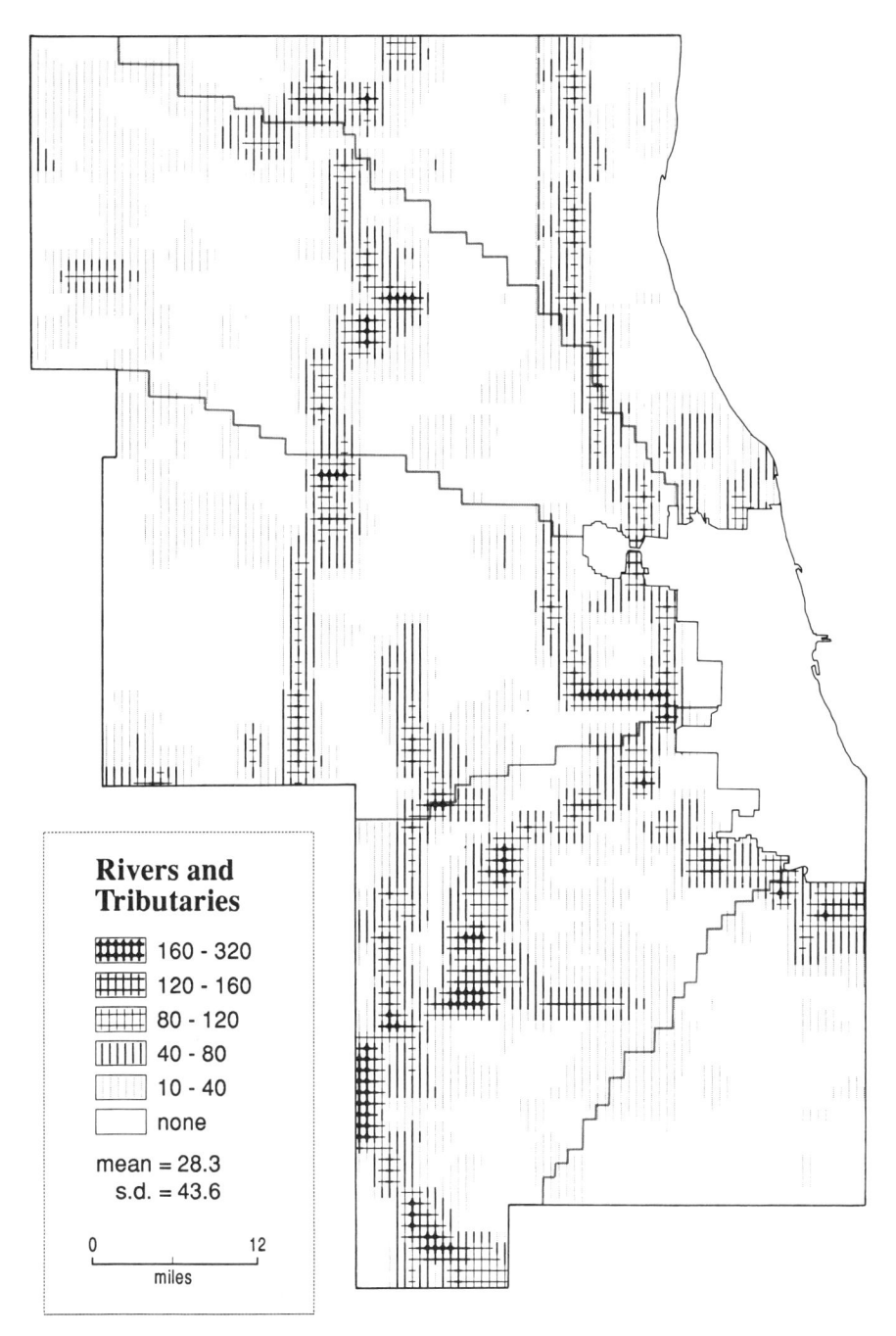

Fig. 7. *The distribution of suburban rivers and tributaries.*

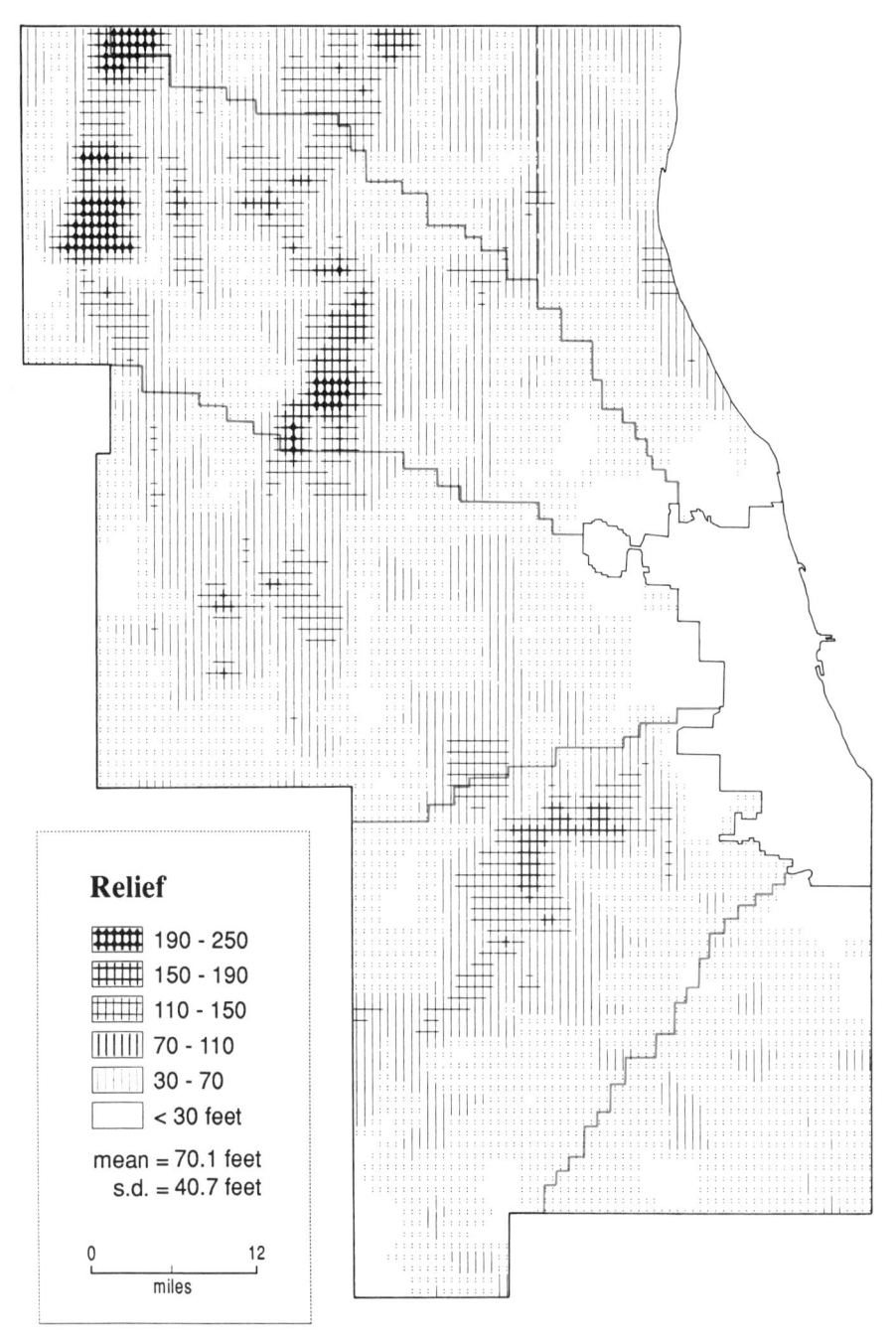

Fig. 8. Relative variation in elevation within designated four-section areas.

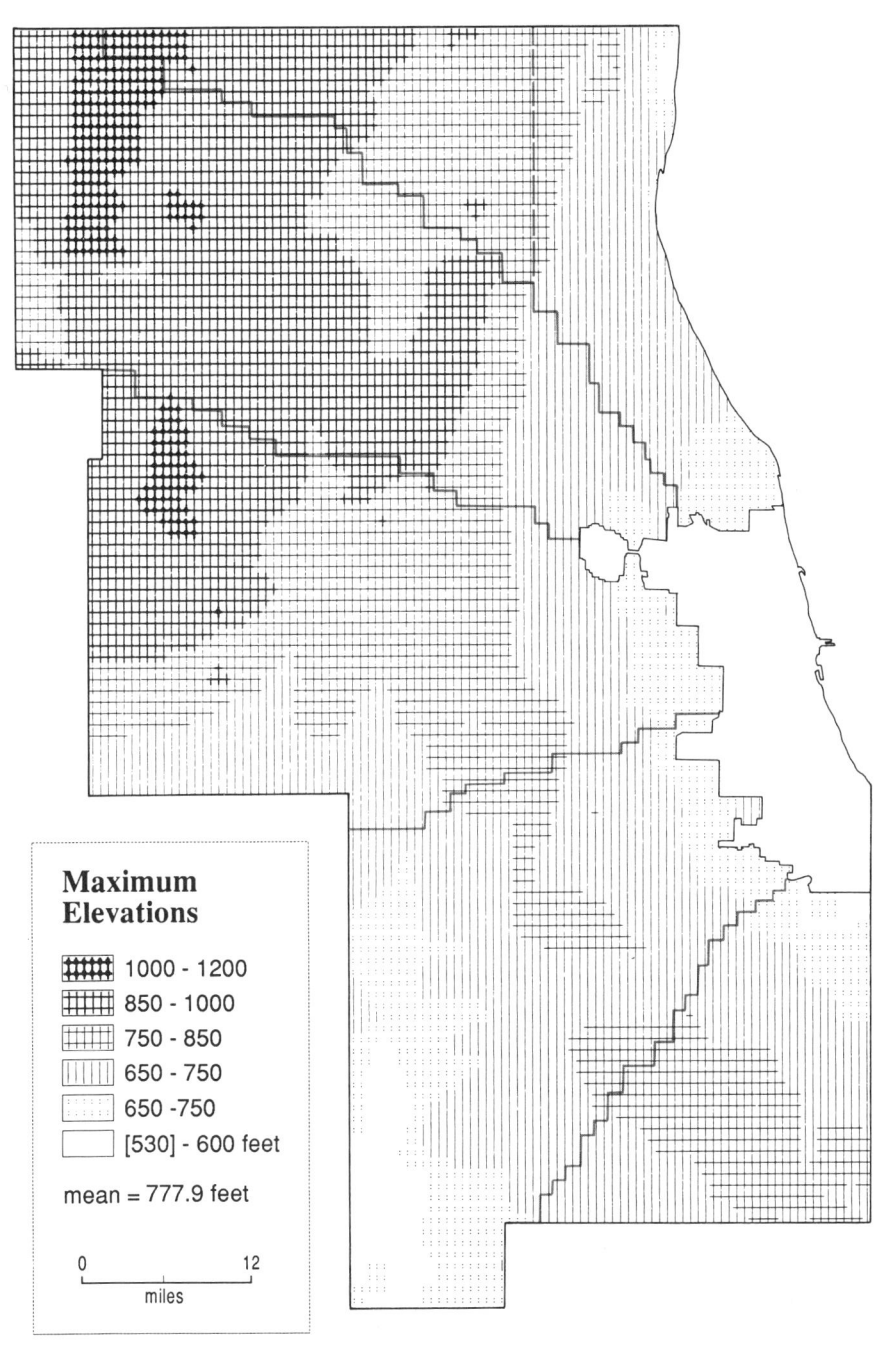

Fig. 9. Maximum elevation within designated four-section areas.

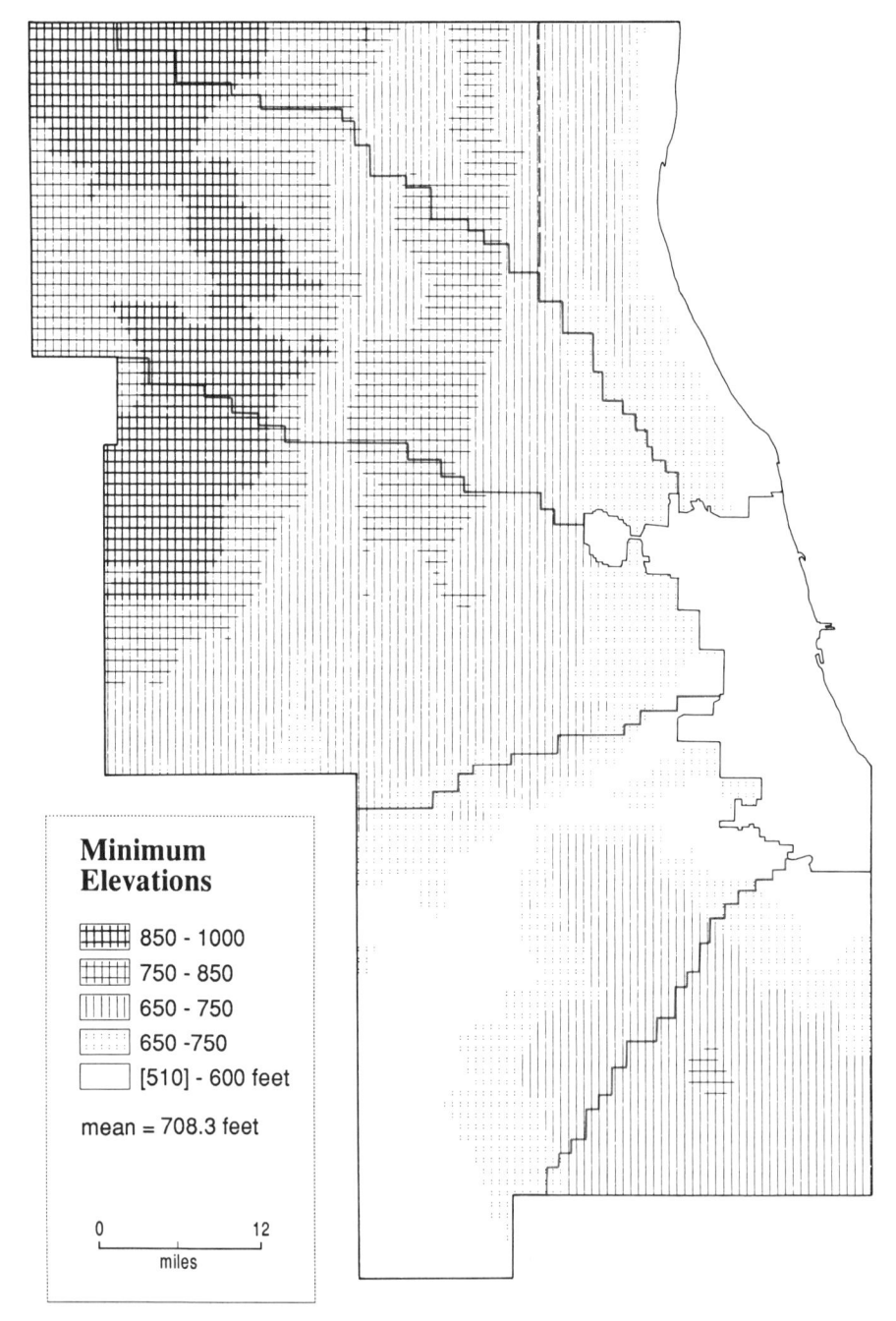

Fig. 10. Minimum elevation within designated four-section areas.

hand, the number of acres does not reveal the number of lakes or the number of golf courses. Nonetheless, the raw totals of each environmental variable were considered a more realistic and practical method of measuring an area's physiographic character and differentiating it from others.

Independent Variables: Intrametropolitan Accessibility Measures

Distance. The point at which the four sections of each aggregate meet became the area's representative centroid. The township-range notation was converted into a decimal-reading Cartesian grid so that the position of each centroid was unambiguously identified by a unique pair of x-y coordinates. The metropolitan region's center is located in Chicago's loop at the intersection of State and Madison streets; it coincides with sectional boundaries. Straight-line distances were calculated between each centroid and the central point. Adjustments were made for the area north of T.41N. in order to account for the nearly one-mile easterly shift of numerically identical range lines. Minor differences in the geometric proportion and alignment of sections were disregarded. The effect of such systematic grid transformation on the calculation of distances was negligible.

Commuter Railroad Access. A circle with a three-mile radius was drawn around each suburban commuter railroad station as marked on NIPC's regional map. An entire four-section aggregate is considered to be within reach if its centroid falls within the circle (figure 11).

Highway Corridor. A corridor extends two miles on both sides of the interstate highways, expressways, or tollways listed on NIPC's regional map; it is not determined by the location of interchanges. The scoring followed the same method as for rail access (figure 12).

The Dependent Variable

Population Density. Population totals of each four-section aggregate for both 1970 and 1980 were divided by the aggregate acreage of usable land. Usable land was defined as the net total of land resulting from subtraction of two major land-use areas. The first incorporated the two standard land-use categories of inland water (Lake Michigan not included) and was recorded by NIPC both in 1970 and 1980 as part of the agency's comprehensive land-use inventories. The second consisted of all areas designated as conservation/ preservation areas or state parks. In contrast to the first group, acreage totals of parks were applied to both years as they came from the ESF file.

The calculation of densities for gradient analysis has evolved two standard approaches. The first proceeds with gross density figures as in the

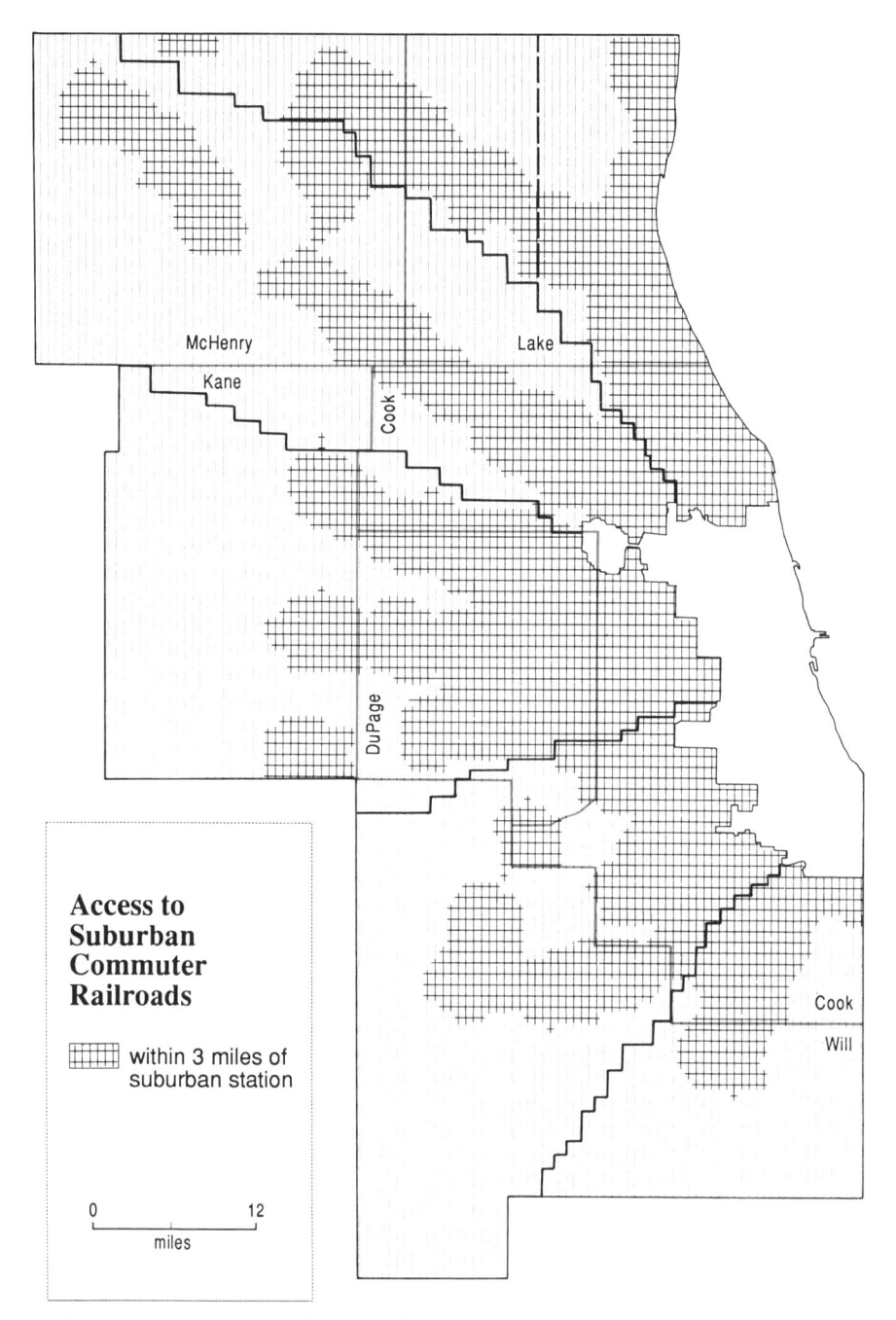

Fig. 11. Centroids of designated four-section areas within three miles of suburban commuter railroad stations.

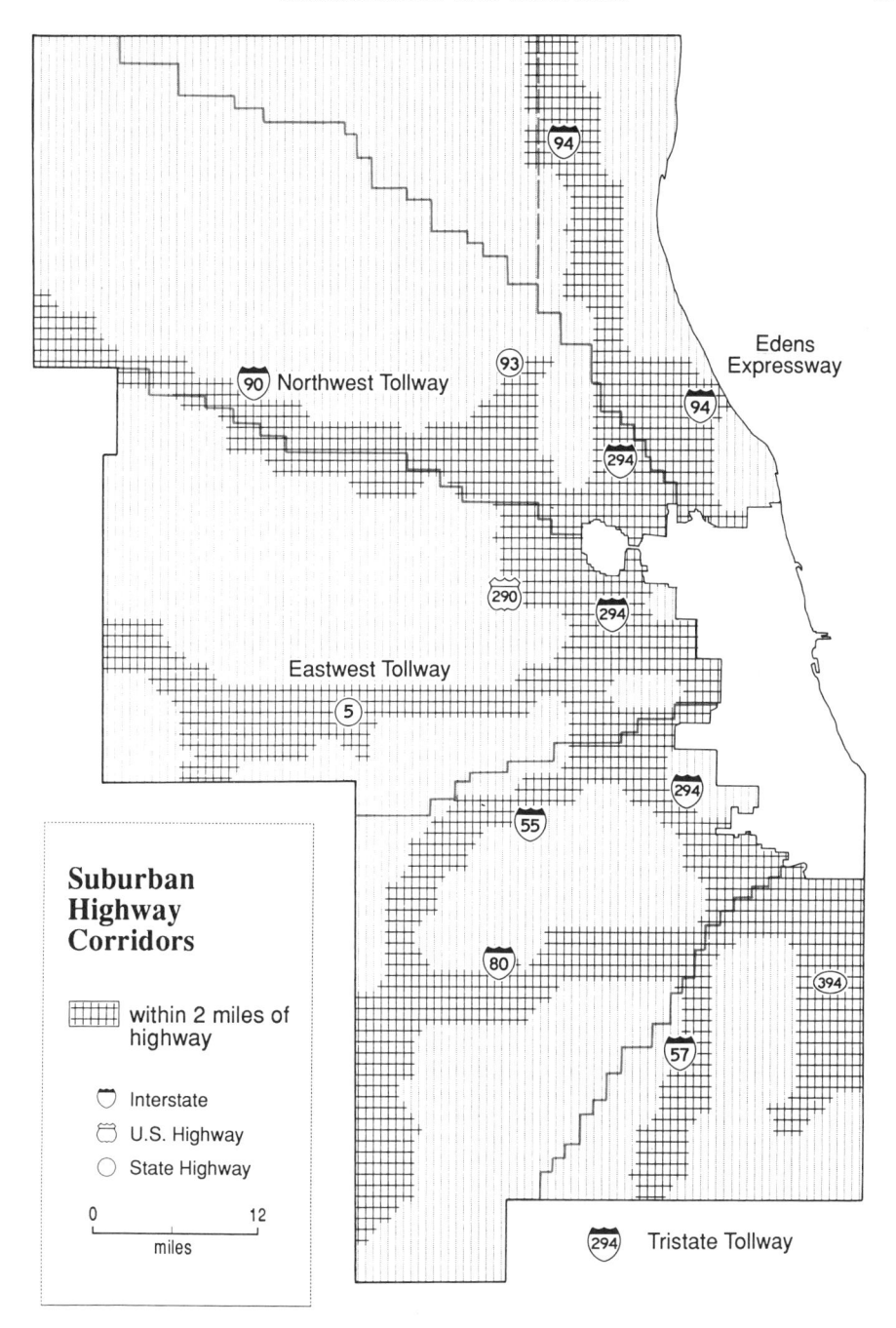

Fig. 12. Centroids of designated four-section areas within two miles of limited-access expressways and interstates.

case of two-point estimates and other derivative ratios based on dwelling density and household size, or actual population counts of municipalities and census tracts. The second employs some form of net density, since some portion of the total available land remains unusable for residential purposes. The decision to use net densities for this study was made for the following reason: environmental conditions are exogenous factors whose level or quantity is location-specific and independent of population density. Assume that there are two areas of identical size both in area and population. Each area contains a lake, with one twice as large as the other. Gross population densities are the same in both areas regardless of the presence of differently sized lakes. After subtracting lake size from total area and calculating net population densities, a comparison of the two areas shows that the one with the larger lake is more densely populated than the other. By basing density measurements on the amount of usable land rather than total land area, a different pattern of spatial association emerges that allows for more incisive inferences concerning the relationship between lake size and population density. A similar argument can be made for excluding public parks and nature areas. Neither land use is initially part of the residential housing and land market. Such reduction in actual land area underlines the principal function of each environmental variable, namely to characterize the physiographic conditions and qualities of a given parcel of suburban land independent of population.

Chapter 5

NATURAL LANDSCAPE AMENITIES AND
POPULATION DENSITIES: SECTORAL PROFILES

> *[H]ow can the the amount of land needed to create a "sense of rurality" be determined? Openness, beauty, wholesomeness are perceptual rather than quantifiable characteristics. The success or failure of the illusion will depend on subjective reactions of the user, not upon a predetermined formula.*
>
> R.H. Platt, *The Open Space Decision Process*

This study's principal thesis holds that natural landscape features play discernible roles in shaping the pattern of suburban population densities. The primary objective of this chapter is to identify the specific landscape attributes and components that are associated with measurable and distinctive shifts in estimated residential density levels. The relative effect of landscape variables is compared to the influence of several of the more conventional measures of urban form, such as linear distance to the central business district and the two intrametropolitan transportation variables of commuter railroad and highway access.

The suburban area of the Chicago metropolitan region is subdivided into five sectors: the north, northwest, west, southwest, and south. In order to account for the specific influence of proximity to Lake Michigan on residential densities, a north-shore margin is carved out of the north sector. Each sector is centered on at least one radial commuter railroad track. Even though the use of sectoral-cum-zonal divisions has greatly diminished as a common strategy of sorting out the complex pattern of urban form and expansion, there are strong reasons for electing such areal division.

On the theoretical side it has been argued that the distended sectoral, rather than pure zonal, pattern of suburban expansion and population densities is typical for the metropolis of the American Midwest.[1] Conceptual models concerned with the general shape of urban areas continue to emphasize the pervasive influence of intrametropolitan transportation on the distribution of urban activities and land uses.[2] Sectorality has been invoked under conditions of transportation constraints and attractive physical environments that are readily exploited to enhance social prestige and provide lush ornamental surroundings.[3] Sectoral divisions may not always be immediately evident or ultimately useful, yet by taking radial transportation corridors as spatial reference markers, sectoral analysis may reveal directional differentials in suburban development and expansion as well as variations in degree and timing. Empirical evidence of population densities around Chicago has consistently demonstrated an irregularly shaped metropolitan area that is distended along established transportation corridors.[4]

On the practical side, the sectoral geometry provides a reasonable spatial division that may assist in pinpointing some of the more local, limited effects of specific natural landscape features. It would be unreasonable to expect that each potentially significant landscape feature should have a sustained regionwide impact on population densities. Moreover, the sectoral division is maintained because it brings together the two principal sets of variables considered in this study: on the one hand there are the qualitatively defined variables of access that are substantially different from distance and direction; this is balanced on the other hand by a set of landscape variables that may further alter the conventionally defined role of distance and add to the qualitative dimensions of suburban amenities.

---

[1] J.S. Adams, "Directional Bias in Intra-Urban Migration," *Economic Geography* 45 (1969): 302-323; J.S. Adams, "Residential Structure of Midwestern Cities," *Annals of the American Association of Geographers* 60 (1970): 37-62 ; K.N. Conzen, "Patterns of Residence in Early Milwaukee," in *The New Urban History*, ed. L.F. Schnore (Princeton: Princeton University Press, 1975), ch. 5; P.O. Muller, *Contemporary Suburban America* (Englewood Cliffs: Prentice Hall, 1981), ch. 2.

[2] L.S. Bourne, ed., *Internal Structure of the City* (New York: Oxford University Press, 1971), pp. 69-131; D. Ward, "The Place of Victorian Cities in Development Approaches to Urbanization," in *The Expanding City*, ed. J. Patten (London: Academic Press, 1983), ch. 12.

[3] H. Hoyt, *The Structure and Growth of Residential Neighborhoods in American Cities* (Washington, D.C.: Government Printing Office, 1939); H. Hoyt, *One Hundred Years of Land Values in Chicago* (Chicago: University of Chicago Press, 1933).

[4] B.J.L. Berry and F.E. Horton, *Geographic Perspectives on Urban Systems* (Englewood Cliffs: Prentice Hall, 1970), ch. 12; B.J.L. Berry et al., *Chicago: Transformation of an Urban System*, Association of American Geographers Comparative Metropolitan Analysis Project (Cambridge, Mass.: Ballinger, 1976).

TABLE 1. Sectoral dimensions and population statistics.

	North	North-shore	North-west	West	South-west[a]	South
Number of units	160	93	227	242	218	97
Distance to CBD (in miles): shortest/ longest	9/65	9/47	12/70	7/53	9/56	16/44
Total area (acres)	334,162	190,955	541,924	568,061	487,941	228,841
Mean unit size (acres)	2,088.5	2,053.3	2,387.3	2,347.4	2,238.3	2,359.2
Population[b] (1,000s)						
1970	687.6	631.0	620.7	1,351.8	668.8	380.3
1980	695.9	611.5	707.1	1,453.4	767.2	405.9
Population change						
Absolute (1,000s)	8.3	-19.5	86.4	101.6	98.4	25.6
Relative (%)	1.2	-3.1	13.9	7.5	14.7	6.7

[a] Not included is the area of the Joliet Arsenal in the southwestern portion of Will County.

[b] According the Bureau of the Census the suburban population totals for the two years were 3,611,000 and 4,097,000, as compared to the NIPC total of 3,709,200 and 4,029,500, respectively.

The sectoral limits are arbitrary and subjective boundaries that follow the sectional lines of the region's township-range survey. As much as their particular course can be disputed, their immediate usefulness is difficult to disregard. The sectors' spatial dimensions and population counts vary considerably (table 1). It should be noted that the mean acreage of the four-section unit remains fairly stable across all sectors except for the north sector and its north-shore margin. The nominal size of a four-section unit is 2,560 acres. Lower averages are generally due to truncated sections along political boundaries, uncorrected surveyor's error, Indian treaty boundaries, and four-section units split along sectoral limits. Incomplete units along the shore of Lake Michigan further depress the average unit size. The uneven distribution of natural landscape features, demonstrated by the series of maps in the previous chapter, remains essentially undisturbed by the sectoral division (table 2).

The discussion of sectoral population densities proceeds in counterclockwise fashion, starting with the north sector and finishing with the south. The tables of coefficients and standard errors are accompanied by scatter graphs plotting the conventional bivariate relationship of density and distance. Simple correlation matrices and supplemental descriptive statistics of the variables can be found in Appendix B.

TABLE 2. Sectoral mean and minimum/maximum values of natural landscape variables.

	North	North-shore	North-west	West	South-west	South
Forest (acres)	276.2	250.9	336.6	246.0	156.2	200.2
minimum/maximum	0/970	0/970	0/1260	0/1160	0/830	0/880
Nature Park (acres)	87.0	118.4	85.6	79.5	125.0	118.8
minimum/maximum	0/1080	0/1060	0/1700	0/970	0/2110	0/1520
Golf (acres)	40.3	64.9	22.2	33.6	18.1	32.4
minimum/maximum	0/400	0/400	0/530	0/490	0/300	0/340
Lake (acres)	93.3	41.1	27.8	5.0	20.1	5.8
minimum/maximum	0/1270	0/490	0/310	0/80	0/600	0/50
River (cum. count)	24.4	28.4	20.7	24.9	43.4	17.4
minimum/maximum	0/120	0/120	0/185	0/202	0/180	2/165
Relief (feet)	70.8	58.2	84.8	64.3	68.1	49.8
minimum/maximum	0/210	0/120	10/250	10/180	0/180	0/100

The North Sector

The north sector stretches from Chicago's city limits along the shore of Lake Michigan to the Wisconsin border at Winthrop Harbor and fans out to the west until it reaches Hebron in McHenry county. Its span is defined by two commuter railroads. The first runs close to the lakeshore; it follows the tracks of the Chicago-Northwestern Railroad, which serviced the string of suburbs from Evanston to Lake Forest as early as the 1860s.[5] Farther inland a second line follows a parallel course until it takes an abrupt turn to the northwest at Libertyville and passes through the picturesque lake district of Lake County. The sector's principal limited-access highways are the Edens Expressway and the Tristate Tollway.

Regression Results and Interpretation

Distance to Chicago's Loop is the first variable to enter the equation for both 1970 and 1980 (tables 3 and 4). In each year it is followed by two environmental variables: lakes and forests in 1970, and lakes and relief in 1980. Of the two transportation variables used in this study, commuter railroad access appears as a significant and positive influence only in 1970. Locations within the four-mile-wide highway ribbon exert a sizable negative influence in both years, although at the end of the decade there is a slight decrease in the level of significance. None of the remaining variables plays any discernible role in the pattern of residential population densities.

---

[5] E. Chamberlin, *Chicago and Its Suburbs* (1874; reprint, New York: Arno Press, 1974), pp. 378-397.

TABLE 3. Determinants of population density in the north sector, 1970: coefficients and (standard errors) of landscape and access variables.

R^2	.568	.634	.671	.694	.704	.709	.711	.713	.713
Constant	10.042	10.049	10.505	9.654	10.254	10.148	10.160	10.141	10.141
Distance	**-.105**	**-.111**	**-.110**	**-.097**	**-.108**	**-.106**	**-.112**	**-.111**	**-.111**
	(.007)	(.007)	(.007)	(.007)	(.009)	(.009)	(.010)	(.010)	(.010)
Lake		**.002**	**.003**	**.003**	**.003**	**.003**	**.003**	**.003**	**.003**
		(.0004)	(.0004)	(.0004)	(.0004)	(.0004)	(.0004)	(.0004)	(.0004)
Forest			**-.002**	**-.002**	**-.002**	**-.002**	**-.002**	**-.002**	**-.002**
			(.0004)	(.0004)	(.0004)	(.0004)	(.0004)	(.0004)	(.0004)
Rail				*.731*	*.625*	*.588*	*.563*	*.582*	*.582*
				(.215)	(.217)	(.217)	(.218)	(.220)	(.221)
Highway					-.499	*-.574*	*-.567*	**-.596**	**-.597**
					(.217)	(.222)	(.222)	(.225)	(.228)
Golf						.018	.017	.016	.016
						(.011)	(.011)	(.011)	(.011)
Relief							.003	.003	.003
							(.003)	(.003)	(.003)
Nature Park								.0004	.0004
								(.0004)	(.0004)
River									.0001
									(.003)

.015 significant at ≥ 99% level of confidence
.015 significant at ≥ 95% level of confidence
.015 < 95% level of confidence

During the decade distance to downtown Chicago has remained a fairly strong negative influence; its coefficient's initial strength and overall stability have suffered only slightly in 1980. However, beyond the twenty-five-mile range the ability of the distance term to summarize density variability is noticeably reduced (figure 13). During the seventies landscape attributes have increased in relative importance, even though their principal impact remains relatively modest. In 1980, the presence of lakes, together with greater relative relief, increases residential population density by a combined 1.1 percent. Ten years earlier the estimated coefficients for both the lake and forest terms nearly canceled each other, leaving a difference of only .5 percent. For both years environmental variables contribute considerably toward reducing the proportion of variance of population density still unaccounted for. In each case, this proportion shrinks by nearly one quarter (23.8% and 23.3% respectively), with lakes producing by far the largest improvement.

The positive effect of lakes is not surprising. As discussed earlier, the presence of water has consistently been ranked high in surveys of residential or general landscape preferences. The recent emergence of relative relief as a significant factor supports related research, which found a combination of views and panoramic enclosures to represent highly valued landscape

TABLE 4: Determinants of population density in the north sector, 1980: coefficients and (standard errors) of landscape and access variables.

R^2	.537	.625	.645	.656	.663	.667	.669	.669	.670
Constant	9.956	9.965	9.882	10.382	10.201	10.277	10.038	10.017	10.015
Distance	**-.099**	**-.105**	**-.120**	**-.128**	**-.124**	**-.123**	**-.119**	**-.118**	**-.118**
	(.007)	(.007)	(.008)	(.009)	(.009)	(.009)	(.010)	(.010)	(.011)
Lake		**.003**	**.003**	**.003**	**.002**	**.003**	**.003**	**.003**	**.003**
		(.0004)	(.0004)	(.0004)	(.0004)	(.0004)	(.0004)	(.0004)	(.0004)
Relief			**.009**	**.008**	**.008**	**.008**	**.008**	**.008**	**.008**
			(.003)	(.003)	(.003)	(.003)	(.003)	(.003)	(.003)
Highway				-.472	-.554	-.493	-.451	-.477	-.480
				(.215)	(.218)	(.222)	(.229)	(.233)	(.235)
Golf					.021	.021	.021	.021	.021
					(.012)	(.012)	(.012)	(.012)	(.012)
Forest						-.001	-.001	-.001	-.001
						(.0004)	(.0004)	(.0004)	(.0004)
Rail							.180	.192	.194
							(.224)	(.225)	(.227)
River								.002	.002
								(.003)	(.003)
Nature Park									.0004
									(.0004)

.015 significant at ≥ 99% level of confidence
.015 significant at ≥ 95% level of confidence
.015 < 95% level of confidence

characteristics. Its moderately high correlation coefficient with linear distance from Chicago's central business district (CBD) contributes to the apparent instability of the distance term. For once disregarding statistical interaction, it is obvious that by 1980 lakes and relief represent a set of landscape features that in addition to distance articulate the spatial distribution of suburban populations north of Chicago.

Why forests should have a significant and negative effect on population densities is not immediately clear. It would be premature, however, to conclude that the negative effect of forests in 1970 is the result of a collective residential choice that rested on a simple comparison between forests, lakes, and topographic diversity and found in favor of lakes and relief. The strong negative effect of forests in this sector can be interpreted in two ways. There are considerable stretches of wooded areas that to a large extent coincide with and adjoin the floodplains of the Des Plaines River and the lower Skokie River. Land along both rivers is effectively removed from the residential housing market. Along the Des Plaines River, forest preserves form an almost unbroken string stretching all the way to the Wisconsin border. Combined with periodically inundated bottomlands, there is a considerable amount of land unattractive, and to some extent unsuitable, for residential development, at least as long as there exist alternative sites. Since both

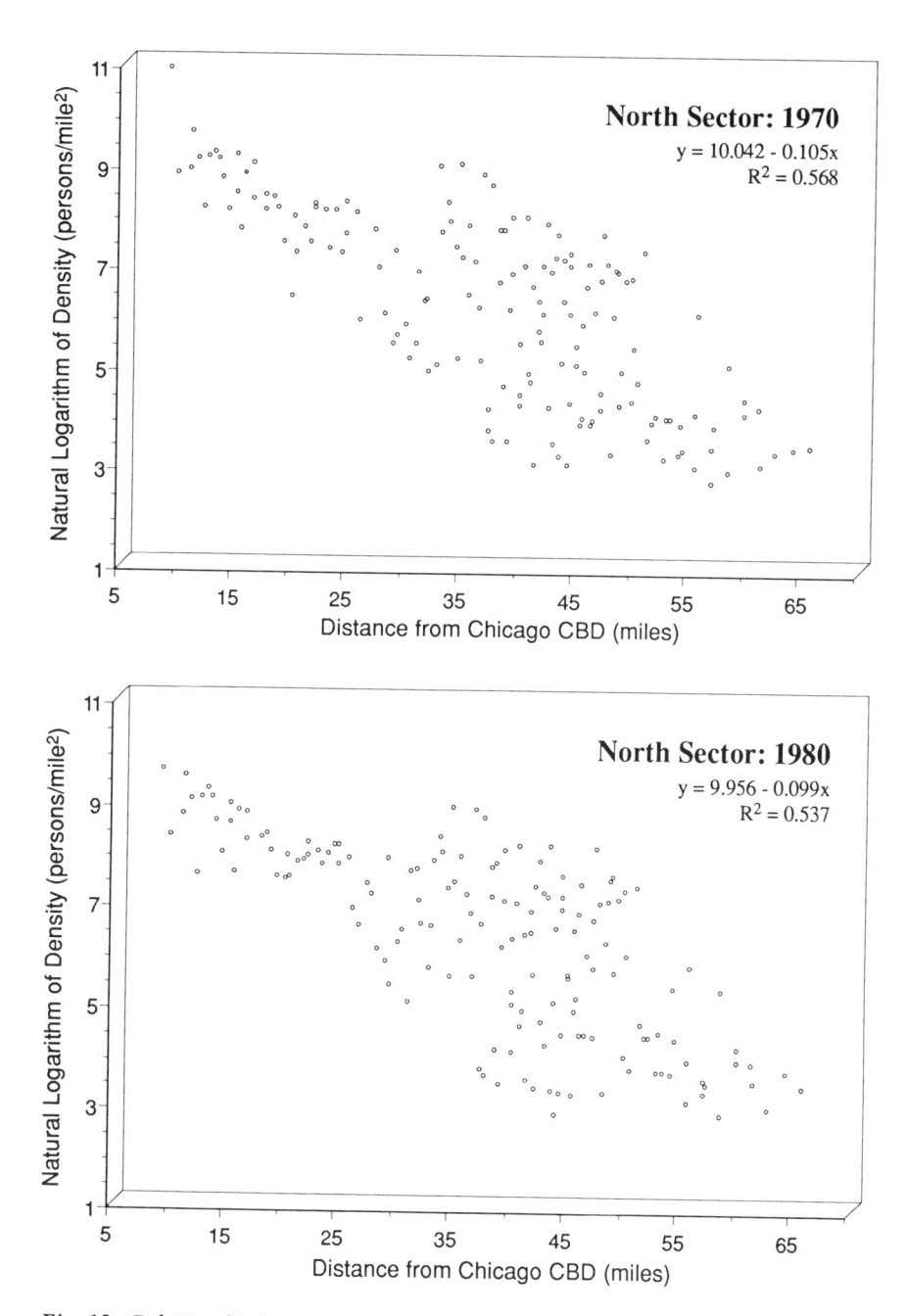

Fig. 13. Relationship between suburban population density and linear distance in the north sector, 1970 and 1980.

rivers follow a radial course, the combined negative effect is felt across all distance bands. Even though the population density figures discount all state parks and preservation areas, the generally negative effect of such restrictive land use on density is shared by the adjoining wooded areas by virtue of their occupying similar topographic and ecological settings. Finally, the formation of natural preservation areas is often possible only when there is little or no pressure created by competing land uses.[6]

Despite the fairly close match between forested areas, protected park lands, and river bottomlands, the negative effect of forests may be the result of an entirely different settlement sequence: in order to preserve the character of wooded areas, lower housing densities are necessary and desirable. This is especially true in the short run; continued residential expansion resulting in denser development would be restricted in order to minimize additional tree removal and long-term damage to trees. Whatever the reason for depressed density levels in 1970 (unattractiveness because of high site development costs in low-lying marshy areas, or attractiveness because of mature trees providing immediate shade and visual enclosures), by the end of the decade estimated population densities in these areas are indistinguishable from density levels at similar distances. Given the scale and time frame of this investigation, there is sufficient evidence that within the northern sector of the Chicago suburban region, forests as such did not exert any sustained, or measurable, influence on the pattern of population distribution.

In 1970 the Tristate Tollway ribbon together with the radially aligned string of forests and nature preserves along the Des Plaines River forms a sharp crease in the population density surface of the northern sector. The steep drop in density along the highway is remarkable; in both years estimated residential densities for areas within two miles of the highway fall by nearly 50 percent. At the end of the decade the crease in the density surface is less pronounced as only the four-mile-wide highway ribbon continues to show depressed density values. At the same time commuter rail access, as the second transportation variable, has become altogether insignificant.

Historically, suburban growth was strongest along the Chicago-Northwestern Railroad and not the Chicago-Milwaukee-St. Paul and Pacific Railroad, which serviced the resort exurbs and country retreats in the hilly lake region on both sides of the Wisconsin border.[7] The cumulative effects

---

[6] For a discussion on how social, economic, and legal factors contribute to the creation of open spaces in suburban areas see R.H. Platt, *The Open Space Decision Process*, University of Chicago Department of Geography Research Paper no. 142 (Chicago: University of Chicago Department of Geography, 1972).

[7] L. Stoneall, *Country Life, City Life* (New York: Praeger, 1983).

of rail access, still noticeable at the beginning of the decade, have been homogenized into a broader regional pattern of population distribution. The principal attraction of the first line has been its proximity to Lake Michigan. In order to separate the effects of Lake Michigan, as the environmental variable, and rail access, as the transportation variable, an additional analysis examined the density surface of the northern lakeshore margin.

The North-shore Margin

For the most part the north-shore margin covers the same area as the north sector with the exception of the northwestern wedge extending into McHenry County. Since its western boundary lies to the west of the Des Plaines River and the four-mile-wide expressway ribbon, the margin covers the diagnosed crease in the density surface. To extend the margin any farther to the west would needlessly dilute the hypothesized effect of distance from Lake Michigan. Measurements employed simple multiples of miles and followed straight-line distances between the fixed center of the standard four-section aggregate and the shoreline.[8]

Regression Results and Interpretation

Distance to Chicago's Loop takes first place in 1970 and in 1980 under virtually identical qualifications, as for the entire north sector; the improvement of 1980 is ever so slight (tables 5 and 6). The sequence of additional significant entries however differs dramatically. Distance to Lake Michigan is a strong second entry for 1970, yet has weakened considerably by 1980. Landscape attributes of any significance are forests in 1970 and relief in 1980. Of the two metropolitan access variables, highway access teeters at the 99 percent level in 1970. In 1980 its significance increases, and it is the second variable to enter the equation. Of all the significant variables, only relative topographic variation is associated with higher population densities.

Although not representing a complete reversal, the decline in importance of distance to Lake Michigan is remarkable. In 1970 the lakeshore term reduces the proportion of unaccounted variation by 30.9 percent; by 1980 it has fallen to a mere 2.8 percent. At the beginning of the decade its effect on density, as yet undiluted by commuter rail access, is more than double that of CBD distance. In 1980 it has fallen below that of the standard distance term and becomes insignificant after the commuter rail term enters the

---

[8] Distance measurements were made on the *Northeastern Illinois Basemap*, 1:126,720, Northeastern Illinois Planning Commission (Chicago: Rand McNally and Co., 1982).

TABLE 5: Determinants of population density along the north-shore margin, 1970: coefficients and (standard errors) of landscape and access variables.

R²	.496	.651	.713	.745	.761	.769	.772	.773	.774	.775
Constant	10.083	10.822	11.268	9.835	10.167	10.098	10.042	9.993	9.922	9.914
Distance to CBD	-.105 (.011)	-.094 (.009)	-.093 (.009)	-.076 (.010)	-.083 (.010)	-.082 (.010)	-.087 (.011)	-.086 (.011)	-.085 (.012)	-.084 (.012)
Distance to Lake Michigan		-.234 (.037)	-.224 (.034)	-.164 (.037)	-.128 (.039)	-.127 (.039)	-.128 (.039)	-.124 (.039)	-.127 (.039)	-.123 (.041)
Forest			-.002 (.0004)	-.002 (.0004)	-.002 (.0004)	-.002 (.0004)	-.002 (.0004)	-.002 (.0004)	-.002 (.0004)	-.002 (.0004)
Rail				.936 (.280)	.854 (.275)	.898 (.274)	.901 (.274)	.891 (.275)	.925 (.281)	.905 (.284)
Highway					-.513 (.216)	-.557 (.215)	-.535 (.216)	-.571 (.223)	-.570 (.224)	-.546 (.228)
Nature Park						.001 (.0004)	.001 (.0004)	.001 (.0004)	.001 (.0004)	.001 (.0004)
Relief							.003 (.003)	.003 (.003)	.003 (.004)	.003 (.004)
Golf								.006 (.009)	.007 (.009)	.008 (.010)
River									.002 (.003)	.002 (.003)
Lake										.001 (.001)

.015 significant at ≥ 99% level of confidence
.015 significant at ≥ 95% level of confidence
.015 < 95% level of confidence

equation. The interaction between these terms is not unexpected. That it is sufficient to remove both variables from the list of significant terms only underlines the extent to which the distribution of population along the northern margin of Lake Michigan has changed as different factors have assumed greater significance.

Based on the two distance variables alone, the 1970 density surface of the north-shore margin tilts simultaneously in two directions (figure 14). It dips gently toward the north away from Chicago and more steeply to the west away from Lake Michigan. During the next ten years the northern slope changes little, whereas the western slope aggrades slightly. The shift in relative directional tilt has been induced by higher densities on the interfluvial rises separating the Skokie and Des Plaines rivers from Lake Michigan. Within the ten-year period the margin's density surface has become unhinged. This does not mean that proximity to the lakeshore has become a less valued landscape characteristic. Nor does it mean that relative relief variation has become a preferred feature. Rather it shows that by 1980 the overall balance in the distribution of population has changed in relative terms. Instead of being concentrated along a fairly narrow strip near the

TABLE 6: Determinants of population density along the north-shore margin, 1980: coefficients and (standard errors) of landscape and access variables.

R^2	.508	.577	.630	.648	.659	.664	.667	.672	.674	.675
Constant	10.083	10.822	11.268	9.834	10.688	10.671	10.718	10.166	10.111	10.058
Distance to CBD	-.101 (.011)	-.108 (.009)	-.128 (.009)	-.123 (.010)	-.125 (.011)	-.125 (.011)	-.125 (.011)	-.118 (.012)	-.117 (.013)	-.116 (.013)
Highway		-.843 (.219)	-.815 (.207)	-.578 (.231)	-.639 (.232)	-.701 (.239)	-.622 (.252)	-.596 (.252)	-.634 (.259)	-.633 (.261)
Relief				.014 (.004)	.014 (.004)	.017 (.004)	.017 (.004)	.017 (.004)	.017 (.004)	.017 (.004)	.017 (.004)
Distance to Lake Michigan					-.086 (.043)	-.102 (.041)	-.103 (.041)	-.105 (.041)	-.086 (.045)	-.081 (.045)	-.083 (.046)
Lake						.002 (.001)	.002 (.001)	.002 (.001)	.002 (.001)	.002 (.001)	.002 (.001)
Nature Park							.001 (.0004)	.001 (.0004)	.001 (.0004)	.001 (.0004)	.001 (.0004)
Forest								.001 (.001)	.001 (.001)	.001 (.001)	.001 (.001)
Rail									.353 (.315)	.340 (.316)	.363 (.324)
Golf										.007 (.011)	.008 (.011)
River											.001 (.003)

.015 significant at ≥ 99% level of confidence
.015 significant at ≥ 95% level of confidence
.015 < 95% level of confidence

shore and within reach of commuter railroads, residential densities have increased farther inland in areas of greater topographic variation. The immediate effect of Lake Michigan remains strongest within a fairly narrow band no wider than two miles (figure 15). However, its indirect effects have been usurped by relief, which now shapes the pattern of residential densities in areas to the west of Lake Michigan.

This interpretation is supported by the dramatic role of highway access with its consistently strong negative effect on densities. Previously supplemented by the negative effect of forests, the highway ribbon at the end of the decade cuts a deep trench into the western edge of the density surface, once more substantiating the existence of a crease that divides the northern sector into two distinct areas of suburban settlement. Estimated densities within the highway ribbon decrease by 84.3 percent. This staggering effect is reduced to 57.8 percent as the remaining significant terms enter the equation. Beyond the twenty-five-mile range the ribbon is defined by IS-94, which up to the Wisconsin border forms a limited-access tollroad. Its widely-spaced exits may have contributed to a more localized suburban settlement pattern. Furthermore, for the last twenty miles in Lake County the

Density (ln) 10, 8, 6, 4

Distance from Chicago CBD →

CBD: Distance from Chicago CBD
DLM: Distance from Lake Michigan

1970 Density = 10.822 − 0.094 $_{CBD}$ − 0.234 $_{DLM}$
1980 Density = 10.452 − 0.096 $_{CBD}$ − 0.132 $_{DLM}$

Miles from Lake Michigan

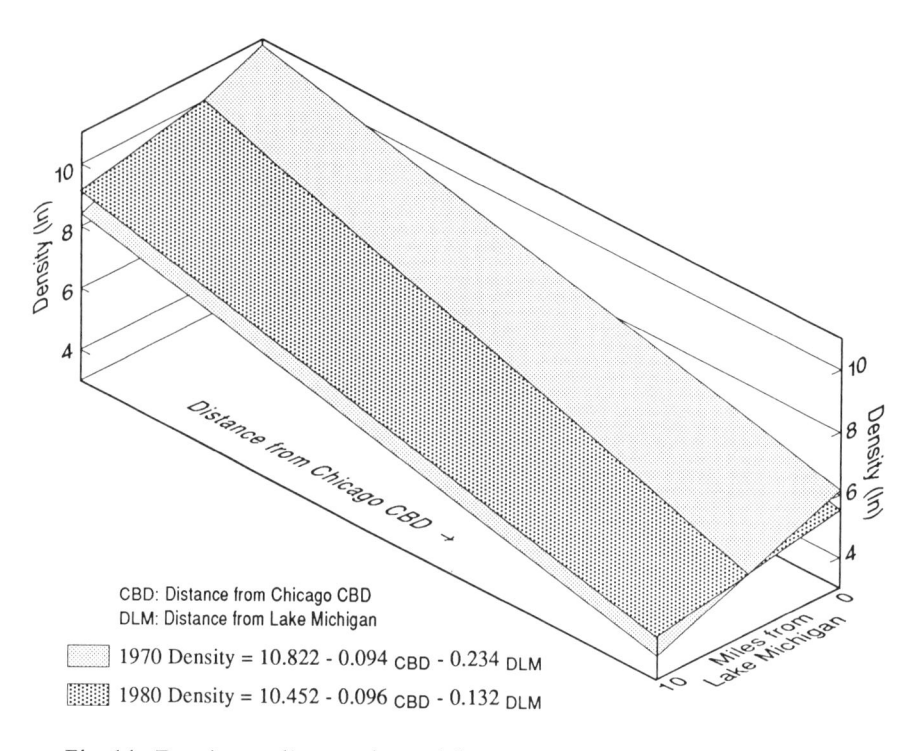

Density (ln) 10, 8, 6, 4

Fig. 14. Density gradient surfaces of the north-shore margin, 1970 and 1980.

highway passes through prime agricultural cropland. Residential development has been restricted to fewer and smaller areas and, by and large, represents a modest expansion of existing suburban centers rather than the explosive growth of scattered subdivisions.

The apparent fold that separates the lakeshore margin from the interior takes on added significance when one considers the respective population figures (table 1). During the seventies the north sector, as a whole, grew by only 1.2 percent. However, taken separately, the north-shore margin lost population, whereas the interior grew. Discounting the margin's portion, the population residing in the rest of the north sector has remained small in number, but it increased by 49.1 percent, compared to a 3.1 percent decrease in the north-shore margin. It is against this background that the patterns of population density and their changes need to be examined.

The crease divides the north sector into two physiographically different areas: the eastern portion promises access to Lake Michigan and moderate topographic variation; the western portion beckons with lakes surrounded by hills. The latter combination seems to have been more successful in attracting new residential development. As for the first pair, stagnating or falling density levels in the older suburbs along Lake Michigan have

North-shore Margin: 1970
$y = 10.083 - 0.105x$
$R^2 = 0.496$

• Within 2 miles of Lake Michigan

North-shore Margin: 1980
$y = 10.083 - 0.101x$
$R^2 = 0.508$

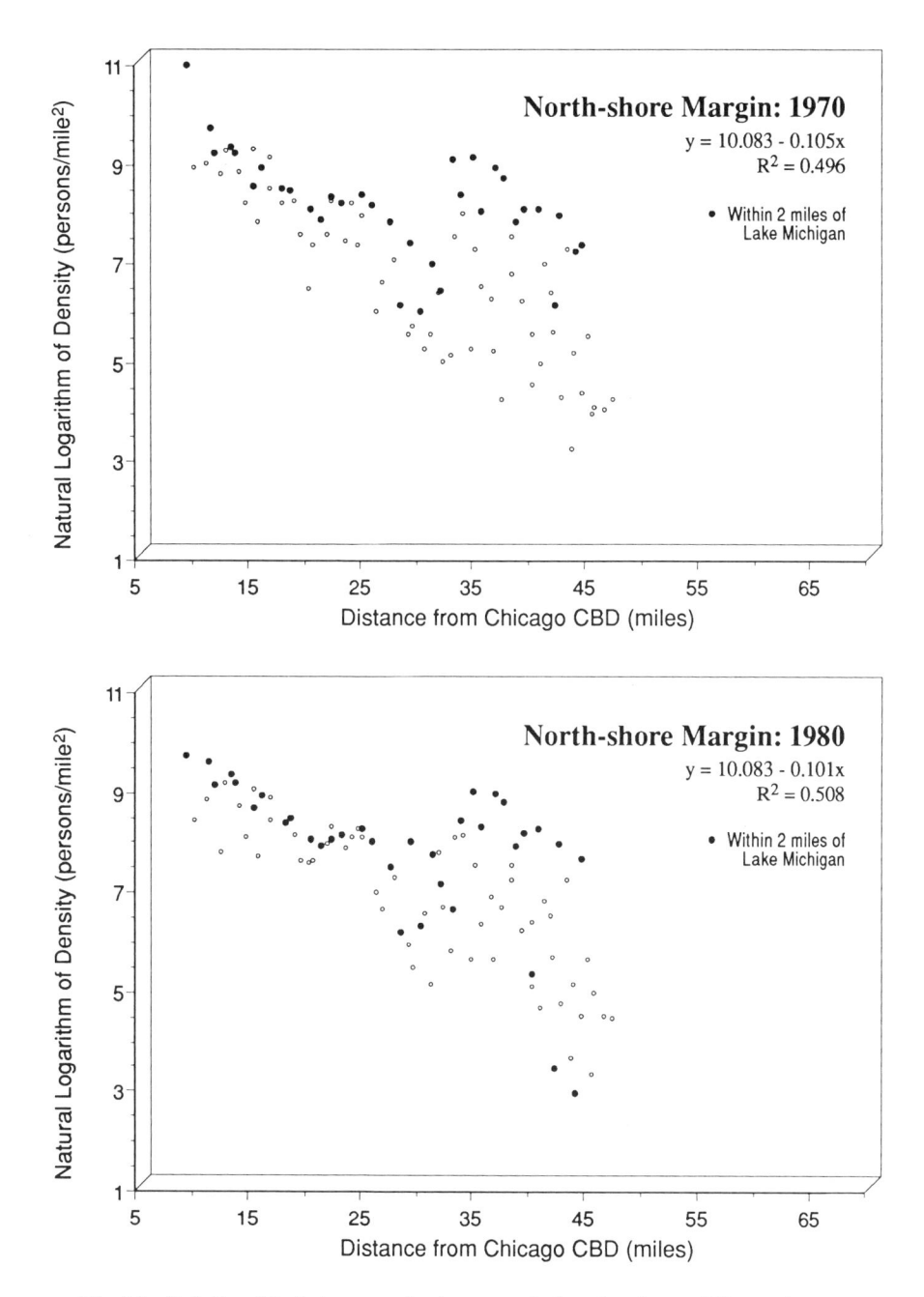

• Within 2 miles of Lake Michigan

Fig. 15. Relationship between suburban population density and linear distance along the north-shore margin, 1970 and 1980.

TABLE 7: Determinants of population density in the northwest sector, 1970: coefficients and (standard errors) of landscape and access variables.

R^2	.621	.682	.696	.708	.716	.721	.723	.724	.727	.727
Constant	9.683	8.867	8.647	8.433	8.286	8.464	8.504	8.319	8.447	8.412
Distance	**-.103**	**-.093**	**-.091**	**-.089**	**-.087**	**-.090**	**-.089**	**-.088**	**-.091**	**-.091**
	(.005)	(.005)	(.005)	(.005)	(.005)	(.005)	(.005)	(.005)	(.006)	(.006)
Rail		**1.004**	**1.033**	**1.028**	**1.025**	**1.008**	**1.018**	**1.042**	**1.026**	**1.030**
		(.155)	(.152)	(.149)	(.148)	(.147)	(.147)	(.147)	(.146)	(.148)
Lake			**.004**	**.004**	**.004**	**.004**	**.004**	**.005**	**.005**	**.005**
			(.001)	(.002)	(.002)	(.001)	(.001)	(.001)	(.001)	(.001)
River				**.006**	**.006**	**.007**	**.008**	**.007**	**.007**	**.007**
				(.002)	(.002)	(.002)	(.002)	(.002)	(.002)	(.002)
Golf					.024	.022	.021	.022	.020	.021
					(.010)	(.010)	(.010)	(.010)	(.010)	(.010)
Nature Park						-.001	-.0004		-.0004	-.0004
						(.0003)	(.0003)		(.0003)	(.0003)
Forest							-.0004	**-.001**	-.001	-.001
							(.0003)	(.0003)	(.0003)	(.0003)
Relief								.002	.002	.002
								(.001)	(.001)	(.001)
Highway										.027
										(.188)

.015 significant at ≥ 99% level of confidence
.015 significant at ≥ 95% level of confidence
.015 < 95% level of confidence

only partially been compensated by slowly rising densities farther inland and at greater distances from Chicago. It is the sector's periphery that is environmentally diversified and attracts most of the recent suburban growth. The consolidated suburban core has exhausted its physiographic potential; here changes in density are brought about by shifting demographic, social, and economic conditions.

The Northwest Sector

The northwestern sector has the greatest radial length of all sectors (tables 7 and 8). Beginning at Park Ridge it stretches beyond Harvard to the extreme northwest corner of the six-county suburban region. It encompasses most of McHenry County as well as the panhandle of Cook County to the west-northwest of Chicago's O'Hare International Airport. The sector is anchored by a single commuter line that follows the tracks of the Chicago-Northwestern Railroad. The first suburbs along this line, such as Arlington

TABLE 8: Determinants of population density in the northwest sector, 1980: coefficients and (standard errors) of landscape and access variables.

R^2	.633	.685	.705	.719	.731	.741	.748	.749	.750
Constant	10.152	9.367	9.100	8.925	8.725	8.825	8.674	8.758	8.575
Distance	-.108	-.098	-.096	-.101	-.098	-.097	-.095	-.097	-.095
	(.005)	(.005)	(.005)	(.005)	(.005)	(.005)	(.005)	(.005)	(.006)
Rail		.967	1.003	1.012	1.005	1.022	1.018	1.007	1.025
		(.160)	(.155)	(.151)	(.149)	(.147)	(.145)	(.147)	(.145)
Lake			.005	.004	.004	.006	.005	.005	.006
			(.001)	(.001)	(.001)	(.001)	(.001)	(.001)	(.001)
Relief				.005	.004	.005	.005	.005	.005
				(.001)	(.001)	(.001)	(.001)	(.001)	(.001)
River					.006	.008	.008	.008	.008
					(.002)	(.002)	(.002)	(.002)	(.002)
Forest						-.001	-.001	-.001	-.001
						(.0002)	(.0003)	(.0003)	(.0003)
Golf							.025	.024	.024
							(.010)	(.010)	(.010)
Nature Park								-.0003	-.0003
								(.0003)	(.0003)
Highway									.144
									(.186)

.015 significant at ≥ 99% level of confidence
.015 significant at ≥ 95% level of confidence
.015 < 95% level of confidence

Heights, were established as early as 1853.[9] In addition, there is a dead-end, two-stop spur north of Crystal Lake running roughly parallel to the upper Fox River valley. The sector's sole radial expressway is the Northwest Tollway (IS-90); it hugs the sector's southern boundary. The sector's throat is traversed by SH-93 and the Tristate Tollway.

Regression Results and Interpretation

In both years linear distance from Chicago's CBD accounts for more then 60 percent of the variance of population density (figure 16). The combined effects of distance and commuter railroad access raise the proportion to more than two-thirds, 68.2 percent in 1970 and 68.5 percent in 1980 (tables 7 and 8). The number of landscape variables that have a significant effect is greater here than in any other sector. By 1980 nature parks and preservation/conservation areas remain as the only insignificant group of landscape features. Ten years earlier this variable together with forests teetered at the

---

[9] Chamberlin, *Chicago and Its Suburbs*, pp. 446-458.

Northwest Sector: 1970
y = 9.683 - 0.103x
$R^2 = 0.621$

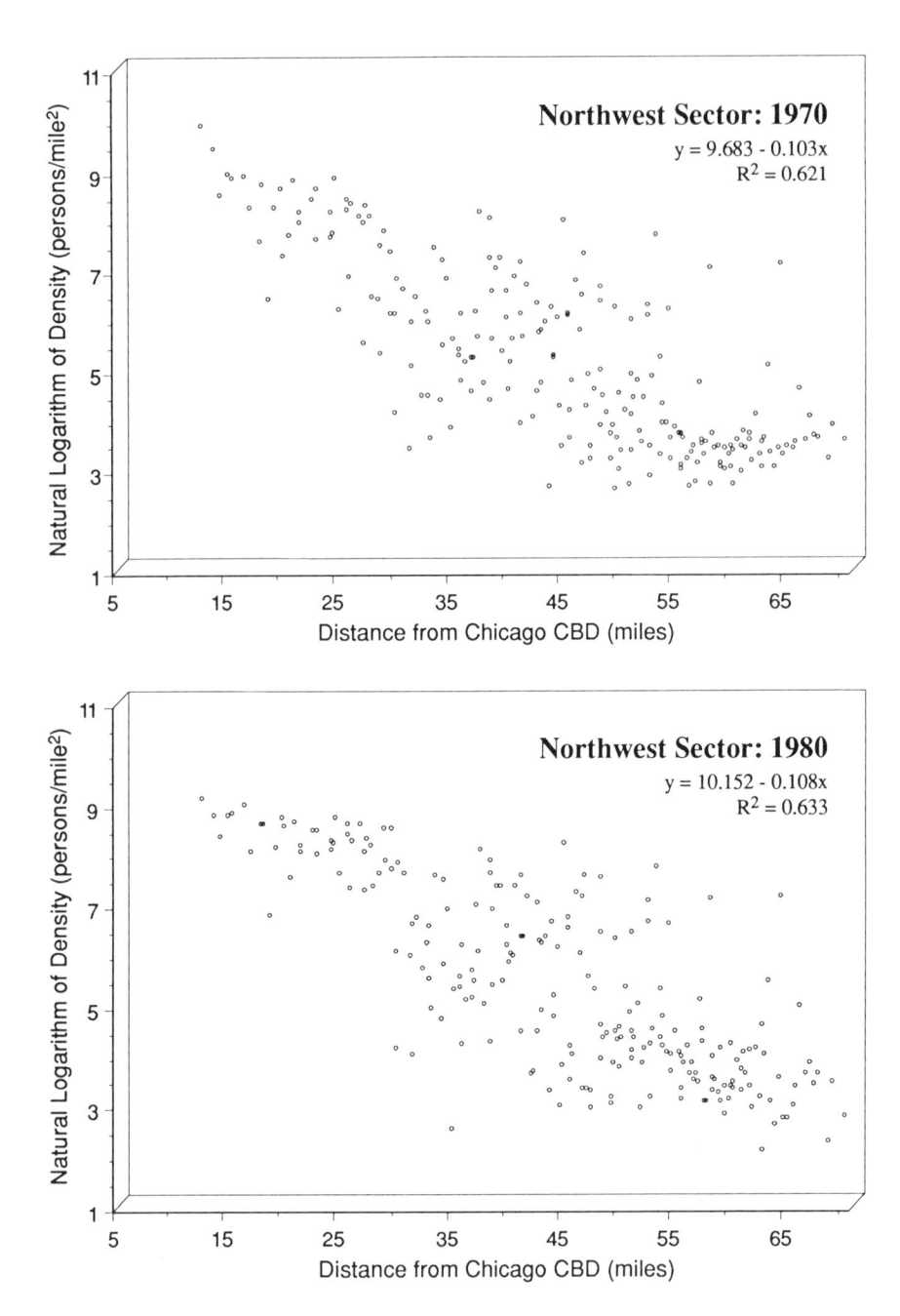

Northwest Sector: 1980
y = 10.152 - 0.108x
$R^2 = 0.633$

Fig. 16. Relationship between suburban population density and linear distance in the northwest sector, 1970 and 1980.

95 percent level; altogether inconsequential at the time was the effect of relative relief. Forests once again have a negative, if trifling, effect; all other significant landscape variables increase population densities. Locations within the four-mile-wide highway ribbons have no measurable impact on densities in either year.

At the beginning of the decade the addition of landscape variables elevates the coefficient of determination by 4.2 percent; in 1980 it jumps by 6.3 percent, reducing by one-fifth the proportion of variance in density initially unaccounted for. For the first time golf courses are a significant factor; in relative terms the presence of golf courses raises residential densities by more than 2 percent. In comparison, the combined effect of the remaining landscape variables never contributes more than 1.7 percent. Both surface water variables, lakes and rivers, sustain a strong effect during the decade. By 1980 the negative coefficient of forests has strengthened. As was the case in the northern sector, the effect of relief shows the most dramatic improvement.

The major portion of lakes and rivers are represented by the upper Fox River valley and the lake district straddling the McHenry-Lake county line from the Wisconsin border to Cook County. Despite a relatively low sectoral mean (tables 1 and 2), lakes are a fairly common feature. Slightly less than one-half (48.5 percent) of the sector's 227 areal units contain at least one ten-acre lake as compared to 54.5 percent in the north sector. Physiographically most of the northern sector's extreme periphery belongs to the same complex of landforms and land cover. As in the north, the presence of lakes has a strong, positive influence on residential densities; at the end of the decade it is joined by relative relief as another landscape feature associated with higher densities. In contrast to the north, the positive effect of lakes and relief is supplemented by rivers. Along the Fox River and its tributaries, which drain most of the area's lakes, are some of the faster growing suburban communities such as Lake Barrington, Oakwood Hills, and Lake in the Hills.[10]

---

[10] The ten-year growth rates for these communities were 562.8, 103.1, and 74.4 percent, respectively. The names mark them unmistakably as more recent suburban creations. Of the seventy-three suburban communities incorporated between 1945 and 1970 nearly a third (31.5 percent) were in the northwest sector. During the 1970s only one of the seven new suburban municipalities was not located in this sector. Residential expansion and growth were not exclusively directed to new locations. First platted in 1836, McHenry is one of the older towns of the region. On the west bank of the Fox River just south of the Chain O'Lakes it offers both rustic charm and access to outdoor recreation. During the seventies it grew by 61.1 percent. Data based on Northeastern Illinois Planning Commission, *Suburban Factbook* (Chicago: Northeastern Illinois Planning Commission, 1971) and *Regional Data Report* (Chicago: Northeastern Illinois Planning Commission, 1978); U.S. Bureau of the Census, *1980 Census of Population*, PC80-1-B15 (Washington, D.C.: Government Printing Office, 1982), table 14.

The progression and pattern of suburban expansion in this part of the metropolitan area seem clear: during the early stages access to water, standing or running, is the primary objective and the reason for discounting distance; in the beginning the majority of new residents can realize this objective. During the later stages as building sites near water and on level land become scarce and expensive, most new residential development is pushed into the surrounding hills. View of the lake may compensate for direct lake access. Finally, residential expansion reaches the forested portions of the hilly countryside with the immediate result of slightly depressed densities. In addition to the apparent relationship between relief and forests there are the lingering edge effects of such restrictive land uses as state parks and nature preservation areas.[11]

Golf courses are throughout the decade a significant variable. As an artificially created and maintained natural landscape complex, they are fairly independent of most other landscape features. Their construction may even ameliorate originally unattractive surface conditions by improving drainage and adding new vegetation. In both 1970 and 1980 their effect on residential densities exceeds the total effect of all the remaining significant landscape variables and even shows a slight increase at the end of the ten-year period. In a highly competitive, rapidly expanding suburban housing market such as the northwest, which is the second fastest growing sector in the region (table 1), developers, builders, buyers, and sellers alike have realized that golf courses can be an invaluable asset. New subdivisions, anchored by a standard golf course, frequently feature mixed residential land uses consisting of single-family homes and apartment-townhouse complexes. Instead of using available land to increase the number of detached home lots to the exclusion of open spaces, planned unit developments preserve open space and provide for public access. The clustered neighborhood design with a mixed pattern of housing types allows recreational open space to be adjacent to each dwelling. Golf courses are ideally suited to fulfill these objectives.[12]

Unfortunately, the environmental data set applies uniformly to both years, so that it is impossible to see whether the number of golf courses in-

---

[11] The intimate association of the three features is illustrated by the case of Lincoln Ridge, Massachusetts, an award-winning planned unit development: 70 percent of the forested land was preserved in its natural state and deeded to perpetual conservation with public access guaranteed. Residential site development sought to balance attractive views and preservation of the existing vegetation. See S. Tomioka and E.M. Tomioka, *Planned Unit Developments: Design and Regional Impact* (New York: Wiley, 1984), p. 16; also see R.S. Doney, B. Evered, and C.M. Kitchen, "Effects of Tree Conservation in the Urbanizing Fringe of Southern Ontario Cities, 1970-1984," *Urban Ecology* 9 (1986): 289-308.

[12] Tomioka and Tomioka, *Planned Unit Developments*.

creased, indirectly confirming such marketing strategies. That the sectoral mean is much lower than that of the north sector, particularly the northshore margin, suggests that factors other than relative size play a determining role (table 2). The absence of a golf-course effect in the north is the result of timing. There golf courses were not an integral element of advancing suburban settlement, inasmuch as the game of golf was not yet a sport to be played by the common person. Construction was left to exclusive country clubs or a few provident community councils and occurred on land not zoned for residential uses. The size of the golf-course effect in the northwest also shows that in relative terms the influence of natural landscape features remains subtle even under the best of circumstances; and the northwest is the best that the region as whole has to offer in physiographic diversity.

Finally, the density ridge along the single commuter railroad shows no signs of erosion. In contrast to the effect of IS-94 in the north sector, the Northwest Tollway remains completely insignificant. The lack of any effect is not explained by the highway's marginal course along the sector's southern flank. The ribbon is wedged between two commuter rail lines and, most important, skirts along the northern edge of the west sector, which is the region's most extensively developed suburban area. As a result the pattern of suburban settlement and expansion was too well established to be disrupted.

The West Sector

The west sector encompasses most of Kane and DuPage counties and the western portion of Cook County. Only slightly larger in area than the northwest, it has almost twice the population (table 1) and ranks first among all sectors in absolute population gains. Its location next to Chicago's narrow waist and a total of three commuter rail lines were the major facilitating factors of its rapid and continued population growth; there is no commuter service beyond the Fox River valley. The Eastwest Tollway as an extension of the Eisenhower Expressway is the sector's only radial expressway. In addition, the Tristate Tollway arcs the city's west side at a distance of about fifteen miles from downtown; U.S. 290 runs from Elmhurst to Schaumburg, where it joins with SH-93.

Regression Results and Interpretation

As was the case in the northwest sector, distance and commuter rail access are the first two variables to enter the equation for the respective years (tables 9 and 10). The rail term's coefficients are much larger than in the two previous sectors and bear witness to the rail-bound pattern of suburban settlements that has dominated residential expansion in this part of

TABLE 9: Determinants of population density in the west sector, 1970: coefficients and (standard errors) of landscape and access variables.

R^2	.688	.770	.780	.786	.789	.792	.794	.795	.795
Constant	10.675	7.615	7.489	7.381	7.391	7.095	7.126	7.084	7.131
Distance	-.148	-.085	-.084	-.089	-.089	-.085	-.086	-.085	-.086
	(.006)	(.009)	(.009)	(.009)	(.009)	(.009)	(.009)	(.009)	(.009)
Rail		2.041	1.939	1.878	1.938	1.967	2.002	2.000	1.993
		(.221)	(.219)	(.218)	(.219)	(.219)	(.219)	(.219)	(.220)
River			.005	.005	.005	.005	.005	.005	.005
			(.002)	(.002)	(.002)	(.002)	(.002)	(.002)	(.002)
Relief				.005	.005	.006	.006	.006	.005
				(.002)	(.002)	(.002)	(.002)	(.002)	(.002)
Lake					-.012	-.012	-.010	-.010	-.009
					(.006)	(.006)	(.006)	(.006)	(.006)
Highway						.244	.243	.259	.261
						(.142)	(.141)	(.142)	(.142)
Golf							-.013	-.014	-.015
							(.009)	(.009)	(.009)
Forest								.0003	.0004
								(.0003)	(.0003)
Nature Park									-.0003
									(.0004)

.015 significant at ≥ 99% level of confidence
.015 significant at ≥ 95% level of confidence
.015 < 95% level of confidence

the region since the middle of the nineteenth century.[13] Rivers and relative relief are the only landscape terms that, after trading places, remain as additional, significant variables at the end of the decade. In 1970 the presence of lakes appears briefly as a negative factor before it becomes insignificant. None of the other variables have any measurable impact on population densities.

The first two variables, distance and commuter rail access, account for well above 70 percent of the variance of density in both years; however, during this period the proportion of accounted variance falls by 4.5 percent. This loss is never recaptured by any or all of the landscape variables, so that the overall effect of environmental variables actually declines. In relative terms the influence of river locations is less pronounced in 1980 than in 1970. The coefficient of relief, on the other hand, has grown in size and strength. The short-lived negative impact of lakes may be related to size as

---

[13] C. Abbott, "Necessary Adjuncts to Its Growth: The Railroad Suburbs of Chicago, 1854-1875," *Journal of the Illinois Historical Society* 73 (1980): 117-131.

TABLE 10: Determinants of population density in the west sector, 1980: coefficients and (standard errors) of landscape and access variables.

R^2	.653	.725	.741	.747	.747	.748	.748	.749	.749
Constant	10.720	7.960	7.767	7.690	7.707	7.707	7.691	7.736	7.713
Distance	**-.139**	**-.082**	**-.090**	**-.089**	**-.090**	**-.090**	**-.090**	**-.091**	**-.091**
	(.007)	(.009)	(.009)	(.009)	(.009)	(.009)	(.009)	(.009)	(.010)
Rail		**1.840**	**1.728**	**1.663**	**1.688**	**1.670**	**1.667**	**1.660**	**1.662**
		(.233)	(.288)	(.229)	(.231)	(.232)	(.233)	(.234)	(.235)
Relief			**.008**	**.008**	**.008**	**.008**	**.007**	**.007**	**.007**
			(.002)	(.002)	(.002)	(.002)	(.002)	(.002)	(.002)
River				.004	.004	.004	.004	.004	.004
				(.002)	(.002)	(.002)	(.002)	(.002)	(.002)
Golf					-.007	-.008	-.009	-.010	-.010
					(.009)	(.009)	(.009)	(.009)	(.010)
Lake						.005	.005	.006	.006
						(.006)	(.006)	(.007)	(.007)
Forest							.0002	.0003	.0003
							(.0003)	(.0003)	(.0003)
Nature Park								-.0003	-.0003
								(.0004)	(.0004)
Highway									.019
									(.151)

.015 significant at ≥ 99% level of confidence
.015 significant at ≥ 95% level of confidence
.015 < 95% level of confidence

well as it may reflect their origin as water-filled shallow depressions or depleted sand and gravel pits.[14]

The Des Plaines and Fox rivers are the two principal waterways traversing the sector. Both have played major, if distinct, roles in the history and pattern of settlement. Located along the Fox River valley are a string of cities that in 1887 were linked by the Chicago Outer Belt Line. Aurora and Elgin represent two such satellite cities whose early fortunes, though intimately linked to Chicago's growth and vitality, were not typically suburban. As factory towns, both grew into industrial and commercial centers that fulfilled economic functions not shared by the model suburb of daily commuters to the central city, such as Riverside, a precursor of planned unit development of classic proportions. Located at the Des Plaines River, its curvilinear street plan and elevated residential sites accommodate and highlight the area's rolling topography. Public parks, unfenced common greens, and a

---

[14] The mean value of five acres is the second lowest in the region. The maximum size is only eighty acres (table 2). The reader may want to speculate on the apparent interaction between the golf course and lake terms and the sudden switch in sign of the latter, intimating that golf courses have had a mitigating effect.

spilldam across the river to control the flow of water and allow pleasure boating all were elements that supported and enhanced the community's rustic appeal. Supplemented by planted vegetation, Riverside's picturesque surroundings became the hallmark of the "romantic suburb."[15]

That population densities are higher near rivers needs therefore to be examined in light of the sector's historic development. The Fox River was primarily a resource supporting industrial growth.[16] The Des Plaines River, on the other hand, functioned as an amenity supporting the recreational outdoor activities of a growing suburban population. The spatial gap between these two patterns of settlement is closing. By 1980 the broad front of suburban development to the west of Chicago began to spill over the older urban areas of the Fox River valley and their suburban extensions. The sharp break in residential densities at the 30-35 mile range is a feature soon to be erased by the advancing wave of suburban expansion (figure 17).

As the relative importance of river location is dissipating, topographic variation assumes greater significance. To understand why relative relief has gained in importance, several factors need to be considered. First, the sector continues to experience dramatic increases on top of an already large population base. Nearly one third of the region's total growth occurred in the west sector (table 1). Second, the sector is the most extensively, if not densely, settled area of the entire suburban region. In this context, then, it is difficult to argue that relief has become a valued site characteristic out of choice. Much of the land still available for new residential construction is in areas away from rivers and with marginally greater local relief. Considering the sector's population size and long settlement history, it is likely that previous suburban expansion had already occupied the more desirable and environmentally attractive sites that combined both landscape elements.[17]

---

[15] K.T. Jackson, *Crabgrass Frontier: The Suburbanization of the United States* (New York: Oxford University Press, 1985), pp. 79-81. Also see J.G. Fabos et al., *Frederick Law Olmsted, Sr.* (Amherst: University of Massachusetts Press, 1968).

[16] This does not mean that the Fox River did not offer scenic locations that attracted residential development. The important difference to keep in mind is that the early growth of Aurora and Elgin was not primarily the result of offering pleasant physiographic surroundings to prospective suburban residents. As the western reaches of recent suburban expansion engulf the Fox River valley, such attractive settings have become valued locational characteristics. What their effect will be on the pattern of population density is uncertain. Given the time frame and method of analysis, it is impossible to pry apart the different density levels, one brought about by earlier industrial and commercial growth, the other associated with recent suburban expansion.

[17] Olmsted and Vaux, who masterminded Riverside's comprehensive site plan and design, determined that the location was "the only available ground near Chicago which does not present disadvantages of an almost hopeless character." Reported by Jackson, *Crabgrass Frontier*, p. 80.

Fig. 17. Relationship between suburban population density and linear distance in the west sector, 1970 and 1980.

The apparent decline in the importance of rivers is similar to the trend observed in the north-shore margin, where the previously strong effect of distance to Lake Michigan diminished and was replaced by relief. In the west as in the north the patterns of population density of the suburban core area close to the city are undergoing changes that appear to be, if not independent of, than at least overriding the historic significance of attractive physiographic conditions. This conclusion does not negate the value of comparing population densities to variations in natural amenity endowment. Admittedly, looking at aggregate population density as a diagnostic feature compresses multiple layers of suburban expansion with different time depth, geographic reach, and intensity into a smooth surface of beguiling simplicity. Inasmuch as this method of analysis succeeds in identifying the outward appearance of settlement patterns, it generally fails to separate them according to their modern or historic origins.

The Southwest Sector

The southwest is the region's fastest growing sector and ranks a close second in absolute gains behind the west (table 1). Its transportation network displays a distinctive pincerlike geometry not duplicated in any of the other sectors. Two commuter rail lines as well as two interstate highways follow the sector's flanks before converging in or near Joliet. For both commuter lines Joliet is the final destination. A third commuter railroad following a central course never reaches the city, as it terminates at Orland Park. Since there are no commuter railroads servicing the area beyond Joliet, it is expressway IS-55 that facilitates access to the sector's extreme southwest corner. In contrast to the previous sectors, all expressways leaving Chicago's southwest side remain toll-free roads. This does not apply to the Tristate Tollway, which continues its circular course around Chicago.

Regression Results and Interpretation

The 1970 sequence of entries runs counter to all previous sectoral profiles (table 11). For the first time access to commuter rail service is better able to account for the variation in population density than linear distance. Forcing distance from downtown Chicago to enter first place yielded a slightly lower coefficient of determination (figure 18). Not only is access to commuter rail a more significant factor than distance, but areas within the four-mile-wide highway corridor also have a strong positive influence on residential densities. At the end of the decade distance moves to its accustomed position, followed by rivers, highways, and commuter railroads (table 12). Of the four significant variables only one, rivers, pertains to a concrete landscape feature. In 1970 forested areas and protected nature parks together with rivers complete the list of significant variables.

TABLE 11: Determinants of population density in the southwest sector, 1970: coefficients and (standard errors) of landscape and access variables.

R^2	.495	.575	.629	.651	.665	.673	.677	.679	.679
Constant	4.150	6.885	6.378	6.616	7.139	7.158	7.147	1.459	1.459
Rail	3.091	1.914	1.749	1.550	1.533	1.481	1.441	1.459	1.459
	(.215)	(.272)	(.256)	(.255)	(.251)	(.249)	(.250)	(.250)	(.251)
Distance		-.072	-.068	-.079	-.089	-.089	-.088	-.086	-.086
		(.011)	(.011)	(.011)	(.011)	(.011)	(.011)	(.011)	(.011)
Highway			.976	.848	.787	.818	.869	.876	.875
			(.176)	(.175)	(.174)	(.172)	(.175)	(.175)	(.176)
River				.006	.007	.007	.007	.007	.007
				(.002)	(.002)	(.002)	(.002)	(.002)	(.002)
Forest					-.001	-.002	-.002	-.002	-.002
					(.001)	(.001)	(.001)	(.001)	(.001)
Nature Park						.001	.001	.001	.001
						(.0003)	(.0003)	(.0003)	(.0003)
Lake							-.002	-.002	-.002
							(.001)	(.001)	(.001)
Relief								.003	.003
								(.002)	(.002)
Golf									-.001
									(.016)

.015 significant at ≥ 99% level of confidence
.015 significant at ≥ 95% level of confidence
.015 < 95% level of confidence

The sector's prominent physiographic features are the Des Plaines and Du Page rivers and two artificial waterways: the Chicago Sanitary and Ship Canal and the Calumet Sag Channel. The Chicago Sanitary and Ship Canal reinforced the boundary effect of the Des Plaines River. The Calumet Sag Channel created a similar, though much subdued, obstacle. The small number of bridges combined with the region's largest forest preserve and park area located at the intersection of the two canals, plus a series of smaller wildlife and forest preserves, virtually blocked off the sector's central, midrange area. Much of the earlier suburban growth was pushed to the margins and spilled into the neighboring sectors; the southwest sector became the orphan of suburban development in the Chicago metropolitan region.[18]

---

[18] As with any ad hoc regional subdivision there exists the possibility of observing spatial relationships between variables that may well be artifacts created by the very subdivision that they are supposed to explain as independent variables. Obviously sectoral boundaries reflected the course of commuter railroads. However, to redraw sectoral boundaries would not change the fact that the central area remained devoid of the kind of suburban development that typically takes place at such distance.

Fig. 18. Relationship between suburban population density and linear distance in the southwest sector, 1970 and 1980.

TABLE 12: Determinants of population density in the southwest sector, 1980: coefficients and standard errors of landscape and access variables.

R^2	.485	.546	.582	.585	.610	.611	.614	.614	.614	.614
Constant	9.633	9.440	7.963	7.338	7.569	7.744	7.766	7.767	7.704	7.696
Distance	-.126	-.132	-.098	-.082	-.093	-.096	-.097	-.097	-.096	-.096
	(.009)	(.008)	(.011)	(.011)	(.011)	(.012)	(.012)	(.012)	(.012)	(.012)
River		.009	.007		.006	.007	.007	.006	.006	.006
		(.002)	(.002)		(.002)	(.002)	(.002)	(.002)	(.002)	(.002)
Rail			1.173	1.308	1.102	1.097	1.073	1.081	1.086	1.087
			(.275)	(.268)	(.266)	(.266)	(.267)	(.269)	(.270)	(.270)
Highway				.849	.715	.694	.699	.690	.692	.694
				(.182)	(.180)	(.182)	(.182)	(.185)	(.185)	(.187)
Forest						-.0004	-.001	-.001	-.001	-.001
						(.001)	(.001)	(.001)	(.001)	(.001)
Nature Park							.0003	.0003	.0003	.0003
							(.0003)	(.0003)	(.0003)	(.0003)
Lake								.0004	.0004	.0004
								(.001)	(.001)	(.001)
Relief									.001	.001
									(.003)	(.003)
Golf										.002
										(.017)

.015 significant at ≥ 99% level of confidence
.015 significant at ≥ 95% level of confidence
.015 < 95% level of confidence

As the distinctive geometry of the sector's transportation network retarded growth in the the sector's midsection, Joliet prospered. It was the destination of two commuter rail lines and, like Aurora and Elgin, it was part of the Chicago Outer Belt Line. Its location on the Des Plaines River just south of Lockport, where the Chicago Sanitary and Ship Canal joins the river, further boosted its status as the industrial and commercial center of the area. From Joliet to Chicago the narrow band of river, canal, and railroad provided the lifeline for new settlements. Some of them, such as Lockport and Lemont, could hardly be considered suburbs, as they grew out of labor camps during canal construction to become service stations for canal operations and maintenance; industrial plants and railroad depots completed their functional base.[19]

---

[19] Lemont's past reaches into the 1840s when the canal's precursor, the Illinois and Michigan Canal, was constructed. More on the settlement history of this area can be gleaned from M.P. Conzen and M.J. Morales, eds., *Settling the Upper Illinois Valley: Patterns of Change in the I&M Canal Corridor: 1830-1900,* University of Chicago Committee on Geographical Studies, Studies on the I&M Canal Corridor no. 3 (Chicago: University of Chicago Committee on Geographical Studies, 1989).

In 1970 the predominantly linear pattern of settlement is still relatively undisturbed. It is accentuated by three linear features: railroads, highway corridors, and rivers. The surprisingly strong positive influence of highway access is the result of the sector's particular settlement history. In other parts of the metropolitan region highways are either insignificant or negative: insignificant because a broadly based, consolidated pattern of suburban settlement is too well established for highways to make a measurable difference, such as in the west; negative because residential development is moving away from highway corridors into more distant, environmentally superior areas, such as in the north.

Highway access is a positive determinant of suburban densities in the southwest for two reasons. First, highways make accessible areas that had been beyond the reach of daily rail commuting. Given the sector's predominantly linear settlement pattern, there is still plenty of space available for residential development at short- to medium-range distances. One branch of the obliquely placed, Y-shaped highway corridors, IS-80 slants across the sector, effectively piercing the vacuum that existed at the center. Second, except for rivers, the physiographic appearance of areas outside the corridor is generally too bland and unappealing to draw additional population.[20] The short-lived effects of nature parks and forests underscore the widespread absence of physiographic diversity. In fact, the greatest diversity featuring extremely hilly (by sectoral standards) terrain, dense forests, and numerous lakes is concentrated in the extensive forest preserves and wildlife preservation areas on both sides of the Calumet Sag Channel. The suburban communities of Palos Hills and Orland Park have successfully exploited their proximity to these areas.[21] Lower densities in forested areas may be the result of either the first stages of residential infiltration or concerted efforts to preserve the natural character of the area. Owing to increased densities throughout the midrange distance bands, neither effect remains significant by the end of the decade.

Rivers are the only landscape feature that remain significant. Their relative effect stays unchanged; however, they improve considerably the level of accounted variation. Not discounting the lingering contribution of Joliet and the string of industrial canal towns near the Des Plaines River,

---

[20] Bolingbrook may here serve as a representative example. Located just north of IS-55, its nominal distance from downtown Chicago ranges from twenty-four to twenty eight miles. Incorporated in 1965, its first residential subdivisions rose among cornfields. Its population grew by 332 percent between 1970 and 1980.

[21] Their respective growth rates during the decade were 152.2 and 260.6 percent.

new residential development has branched out along its tributaries.[22] By the end of the decade the effect of commuter rail access has been reduced to nearly two-thirds of its previous size. The relative change in population densities near highways is similar in size for both years. As distance to downtown Chicago moves to its accustomed position, it appears that the pattern of population distribution in the southwest sector is in transition. Not only have recent changes in population begun to dissolve the historic lineaments of settlement, but a unique order of suburban and exurban development has emerged. It mixes both conventional and modern elements of metropolitan expansion. Infilling at short-to-medium range conforms to the conventional hold of linear distance. However, as the speed of highway travel stretches the envelop of distance, previously remote areas are made accessible to a greater number of potential residents. Highway ribbons and scenic creeks characterize the modern pattern of mixed suburban and exurban development.

The South Sector

Among the five suburban sectors the south ranks last in terms of population and area (table 1). Its radial length of only twenty-eight miles makes it also the shortest sector. Anchored by the Illinois Central railroad the sector contains two radial expressways, IS-57 to the west and IS-94 with its continuation as SH-394 to the east. Whereas IS-57 runs the length of the sector, SH-394 abruptly narrows into a two-lane road before reaching the rural service town of Beecher nearly thirty-eight miles due south of downtown Chicago. Both interstate highways are joined by the final arc of the Tristate Tollway and IS-80. The Indiana state line represents the sector's eastern limit.

Regression Results and Interpretation

The pattern of residential density to the south of Chicago fits the most simple regression model among all sectors (tables 13 and 14). Over the ten-year period the distribution of population in this part of Chicago's suburban region retained a pronounced degree of compaction with little extended peripheral growth (figure 19). Changes in estimated residential density are evenly distributed across all distance bands, resulting in a typically parallel, outward shift of the gradient.[23]

---

[22] The area along Hickory Creek, to the east of Joliet, has experienced a significant population boom. Frankfort is the largest town in the area; it grew by 87.4 percent.

[23] B.J.L. Berry et al., "Urban Populations: Structure and Change," *Geographical Review* 53 (1963): 389-405.

TABLE 13: Determinants of population density in the south sector, 1970: coefficients and (standard errors) of landscape and access variables.

R^2	.763	.780	.788	.792	.801	.804	.805	.806	.807
Constant	13.665	14.775	14.422	14.722	14.423	13.695	14.133	14.290	14.140
Distance	**-.273**	**-.296**	**-.288**	**-.292**	**-.283**	**-.265**	**-.276**	**-.274**	**-.271**
	(.016)	(.017)	(.018)	(.018)	(.018)	(.024)	(.027)	(.028)	(.028)
Highway		**-.680**	-.619	-.610	-.588	-.529	-.531	-.564	-.572
		(.251)	(.252)	(.250)	(.246)	(.251)	(.252)	(.260)	(.261)
Golf			.022	.026	.029	.028	.028	.029	.028
			(.014)	(.014)	(.014)	(.014)	(.014)	(.014)	(.014)
Forest				-.001	-.001	-.001	-.001	-.001	-.001
				(.001)	(.001)	(.001)	(.001)	(.001)	(.001)
Nature Park					.001	.001	.001	.001	.001
					(.0004)	(.0004)	(.0004)	(.0004)	(.0004)
Rail						.365	.342	.348	.396
						(.336)	(.336)	(.338)	(.353)
River							-.004	-.005	-.005
							(.005)	(.005)	(.005)
Relief								-.004	-.004
								(.007)	(.007)
Lake									.005
									(.011)

.015 significant at ≥ 99% level of confidence
.015 significant at ≥ 95% level of confidence
.015 < 95% level of confidence

Distance to downtown Chicago exerts a most overwhelming influence not observed in any of the other sectors. The coefficient of determination even shows a slight improvement in 1980. With nearly 80 percent of the variation in population density accounted for by linear distance alone, it is not entirely unexpected that the influence of the other variables should remain marginal. Highway corridor locations in 1970 and nature parks/preservation areas in 1980 are the only other variables that maintain a consistent level of significance. The former have a sizable, negative impact on 1970 population densities. At the end of the decade, however, highway access has become inconsequential. Instead the presence of forest preserves and nature parks produces a small rise in residential densities.

The extreme compactness of suburban settlement (i.e., a steep density gradient and high central density) is characteristic for both time periods and bespeaks a generally uniform natural landscape that has little to offer in terms of variable topography, surface water, and vegetation cover. Despite excellent access there is no evidence of distended growth along the highway or railroad corridors; there simply is not much there that would encourage

TABLE 14: Determinants of population density in the south sector, 1980: coefficients and (standard errors) of landscape and access variables.

R^2	.798	.807	.815	.820	.823	.826	.827	.827	.828	.828
Constant	13.906	13.542	13.252	13.884	14.419	14.610	14.133	14.187	13.992	14.092
Distance	**-.271**	**-.262**	**-.256**	**-.272**	**-.283**	**-.286**	**-.274**	**-.269**	**-.265**	**-.268**
	(.014)	(.014)	(.015)	(.017)	(.019)	(.019)	(.025)	(.024)	(.025)	(.026)
Nature Park		.001	.001	.001	.001	.001	.001	.001	.001	.001
		(.0003)	(.0003)	(.0003)	(.0003)	(.0004)	(.0004)	(.0003)	(.0003)	(.0004)
Golf			.024	.0222	.020	.022	.022	.023	.021	.022
			(.012)	(.012)	(.012)	(.013)	(.013)	(.013)	(.013)	(.013)
River				-.007	-.007	-.008	-.007	-.008	-.009	-.009
				(.004)	(.004)	(.004)	(.004)	(.005)	(.005)	(.005)
Highway					-.306	-.296	-.260	-.302	-.312	-.302
					(.224)	(.224)	(.230)	(.235)	(.236)	(.238)
Forest						-.001	-.001			-.0003
						(.0003)	(.001)			(.001)
Rail								.227	.267	.330	.295
								(.308)	(.300)	(.314)	(.323)
Relief								-.005	-.005	-.003
								(.006)	(.006)	(.007)
Lake										.007	.007
										(.010)	(.010)

.015 significant at ≥ 99% level of confidence
.015 significant at ≥ 95% level of confidence
.015 insignificant at < 99% level of confidence

suburban settlement at greater distances.[24] The average value or coverage of landscape variables is much lower here than anywhere else in the region. The mean relative relief of less than fifty feet is the lowest of all sectors, including the north-shore margin; the mean values of lakes and rivers are similarly depressed.

At the beginning of the seventies nature parks together with golf courses and forests flirt with the 95 percent significance level; not one of them maintains significance in the final equation. The possible causes of the underlying linkages responsible for the apparent instabilities are not clear, since the variables' simple correlation coefficients do not reveal a distinctive pattern of spatial association. Furthermore, by 1980 the earlier instability of the three terms disappears as the nature park term remains the only significant variable besides distance. In 1970 golf courses and nature parks, as the two positive terms, represent the extremes of the natural

---

[24] The term "cornbelt exurbia" aptly captures the principal landscape character of this part of the region. See M.P. Conzen, "American Cities in Profound Transition: The New City Geography of the 1980s," *Journal of Geography* 82 (1983), p. 98.

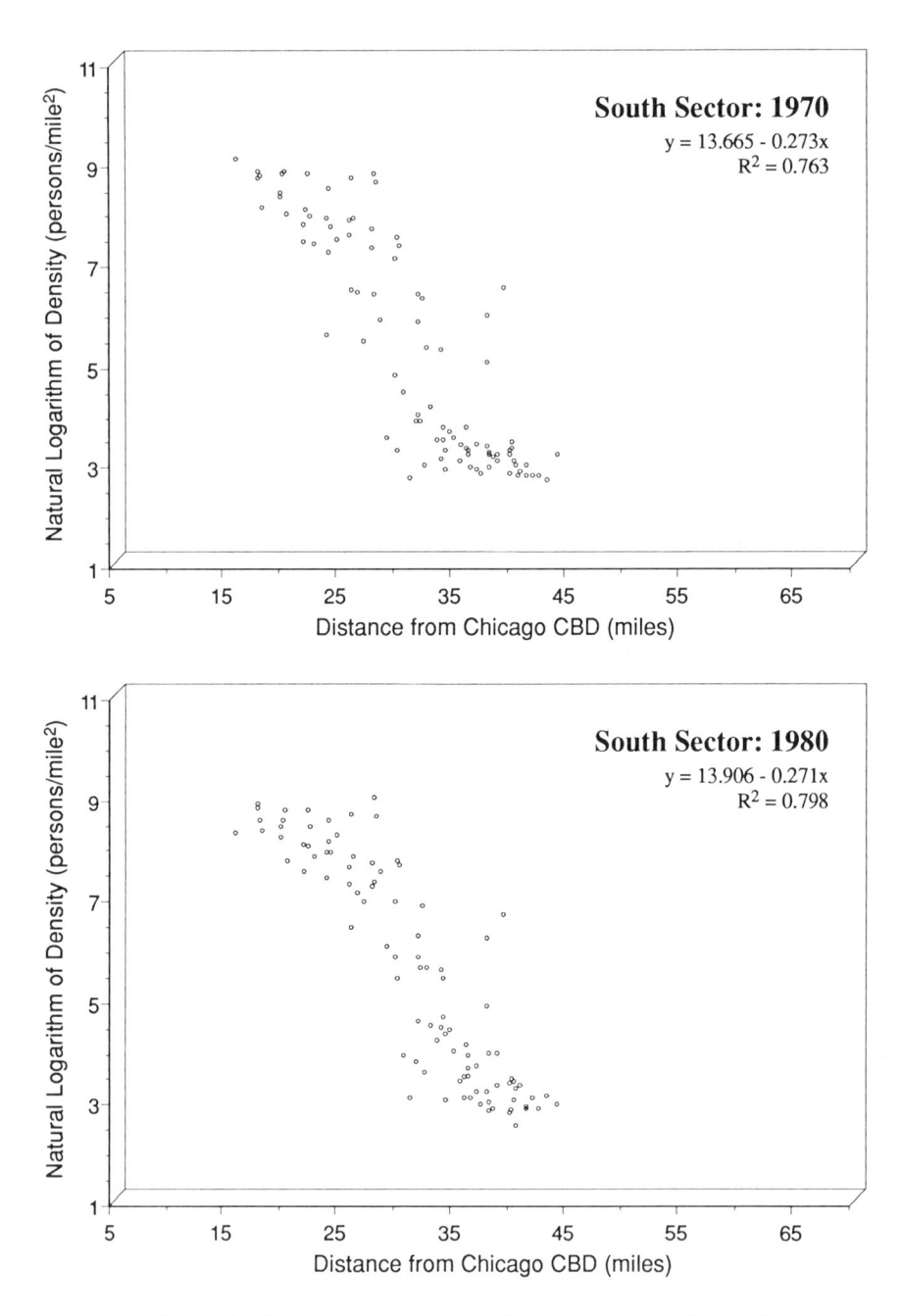

Fig. 19. Relationship between suburban population density and linear distance in the south sector, 1970 and 1980.

landscape spectrum. The first is artificial and can be fully incorporated into a residential development plan. The second is protected as a common good, yet it remains outside the residential housing market. That both have limited influence on residential densities underlines the sector's general plainness in terms of physiographic diversity.

In 1980 even differences in accessibility, by rail or car, play no role in accounting for the distribution of population. The sole commuter railroad running due south terminates in Park Forest South.[25] A second rail line disappears into Indiana and links several exurban lakeshore communities with Chicago. Nonetheless, the lingering effect of rail access still produces a slight decrease in the distance coefficient.

The absence of a larger, fairly distant satellite city further simplifies the settlement pattern in this sector. Although Chicago Heights belongs to the outer ring of industrial cities, such as the old factory towns of Waukegan, Elgin, Aurora, and Joliet, it never achieved the size or status of the others. Much of its initial growth potential was subsequently shared, if not usurped, by the Calumet area to the south and southeast of Chicago, with Gary in Indiana benefiting the most.[26]

To extend the sector to include Indiana's counties of Lake and Porter would not have necessarily enhanced the sector's overall environmental endowment, since it seems likely that the southern margin of Lake Michigan would have to be separated from the rest, as was the case in the northern sector. Within the context of this study, this portion of the Chicago suburban region comes closest to being an isotropic plain, the economist's tabula rasa. At the end of the decade, all traces of apparent differentials in access have been absorbed into a smooth density surface. However, the statistical stranglehold of linear distance to downtown Chicago does not mean that attractive landscape features are completely overlooked in this sector. By the end of the decade forest preserves and park areas retain a marked positive influence on densities; and there are indications that golf courses, as an artificial and managed landscape type, can be a substitute for valued, yet absent natural features. In a physiographically bland landscape, golf courses and parks provide a focus that can function as one of the "natural" set pieces of suburban living.

---

[25] Wanting to erase its image as a mere appendage of Park Forest, this community recently elected to change its name to Governor's Park. Park Forest, an All America City in 1953, was developed in the late 1940s as one of the first comprehensively planned communities in the Chicago area. It gained additional fame through William Whyte's *The Organization Man* (Garden City, N.Y.: Anchor Books, 1956).

26 H.M. Mayer and R.C. Wade, *Chicago: Growth of a Metropolis* (Chicago: University of Chicago Press, 1969), p. 186.

Sectoral Summary and Overview

The preceding series of sectoral profiles has demonstrated that both in 1970 and 1980 physiographic landscape elements had a measurable influence on suburban population densities in the Chicago metropolitan area. The specific effects of landscape features on densities were far from uniform. Features were not equally influential at the beginning or the end of the decade; some persisted, some faded, and others gained in importance or emerged as additional factors. Each sector revealed a unique combination of significant landscape factors. By itself this is not a noteworthy observation, since there exists an undeniable intraregional imbalance in the distribution of potentially attractive natural features. The crucial difference lies not so much in whether a specific feature has a sizable effect, but in when it becomes a significant factor, for how long, and under what conditions.

The bare-boned results of the sectoral analyses highlight two common threads that were hidden within the complex fabric of suburban population densities (table 15). First, the direction of influence of each significant landscape component remains unchanged regardless of sector. The two cases of sign switching may be discounted, as they occur under conditions of pronounced coefficient instability and erratic levels of significance. For the most part, the effects are positive; forests are the only landscape component that has a consistently negative effect on estimated population densities. Second, when considering timing and duration, the list of landscape variables can be reduced to three basic types. Because of the limited time span of this study, this typology is speculative and has to be seen in the context of an arbitrary sectoral subdivision whose boundaries were determined by the course of commuter railroads as the sole criterion.

The first type identifies a group of physiographic features that as landscape fixtures play a fairly consistent role in determining suburban population densities. Lakes and rivers are such fixtures and represent significant factors underlying the pattern of density from the extreme north to the southwestern corner of the region. The attraction of water continues to be a strong incentive for sacrificing ease of intrametropolitan access and overcoming distance. Residential development leaps ahead of the gradually advancing broad suburban front to occupy the choicest sites at the lakeshore or riverbank. Forested areas are representative of the second type of landscape component. Its influence is transitional in character and is strongest during the initial phases of local suburban expansion. The transitional effect of forests can be observed in all sectors except the already extensively developed west sector. Nature areas as potentially attractive alternatives in the absence of general environmental diversity exert a similarly restricted effect. The last type combines landscape features whose influence emerges toward the subsequent phases of suburban consolidation and infilling. Increased

TABLE 15: Significant determinants of sectoral and regional population densities and relative shifts in accounted variation from 1970 to 1980.

	North		North-N-shore		North-N-west		West		South-S-west		South	
	1970	1980	1970	1980	1970	1980	1970	1980	1970	1980	1970	1980
Lake	●	●			●	●	[○]					
River					●	●	●	●	●	●		
Relief		●		●		●	●	●				
Golf					●	●					[●]	
Forest	○		○		[○]	○			○		[○]	
Nature Park					[○]					●	[●]	●
Distance from Lake Michigan			○									
Rail	●		●		●	●	●	●	●	●		
Highway	○	○	○	○					●	●	○	
Distance	○	○	○	○	○	○	○	○	○	○	○	○
R^2 (1970/80 trend)	⇓		⇓		⇑		⇓		⇓		⇑	

● positive coefficient (≥ 95% level)
○ negative coefficient (≥ 95% level)
[ ] unstable confidence level

densities in areas with pronounced relief are an example of this type. Golf courses are a landscape feature whose effects are uneven and cannot readily be categorized. As golf courses become an integral part of residential development and planning, their influence may grow, becoming more widespread and long-lasting.

It is encouraging that none of the landscape variables remained altogether insignificant. This does not mean that the list of environmental features is exhaustive; however the set of landscape components selected for this study is sufficiently comprehensive to identify intraregional differences and their variable effects on suburban population densities. The general variable types were gleaned from an amorphous literature that had no common core of specific research objectives or methodology. There was no presumption as to the kind of effect each landscape component might or should have. Nor were there any prior conditions specifying the possible interaction between the variables. Thus, the results are limited to a particular time and place; they should not be projected indiscriminately into different stages of suburban development, nor are they readily transferable to other metropolitan areas. As will be shown in the next chapter, the typology of

TABLE 16. Changes in the coefficient of determination after adding landscape variables to linear distance and corresponding relative reduction in unaccounted variance of population density.

	1970			1980		
	R^2	R^2	%	R^2	R^2	%
North	.568	.680 (.713)	25.9	.537	.657 (.670)	25.9
North shore	.496	.726 (.775)	45.6	.508	.644 (.675)	27.6
Northwest	.621	.665 (.727)	11.6	.633	.693 (.750)	16.3
West	.688	.721 (.795)	10.6	.653	.694 (.749)	11.8
Southwest	.475	.580 (.679)	20.0	.485	.554 (.614)	13.4
South	.763	.791 (.807)	12.2	.798	.822 (.828)	11.9

NOTE: Previous results of the entire set of variables are in parentheses.

landscape elements fails to reemerge when the suburban region as a whole is analyzed.

Finally, in order to establish a general measure of the relative contribution of all landscape variables toward deciphering the sectoral patterns of population density, separate regression analyses excluded the two intrametropolitan access variables (table 16). Even though the addition of landscape variables raises the coefficient of determination, the reduction in the proportion of unaccounted variance of density varies greatly. A year-by-year comparison shows that by 1980 the overall level of accounted variation has declined in all but two sectors. The only improvement occurs in the northwest and south sectors. However, to conclude that the significance of landscape features is therefore diminishing would be premature. Discounting the special status of the north-shore margin, the relative contribution of landscape variables actually increases in three out of five sectors. It is the importance of intrametropolitan access variables that has diminished and is responsible for the lower overall proportion of accounted variation. The elevated density surface along commuter railroads persists but is less distinctive at the end of the decade. Whenever highway access is significant, its qualitative effect vacillates between negative and positive values.

A transportation-based explanation of suburban population distributions is clearly insufficient to absorb additional distortions of a hypothetically concentric density pattern. Physiographic variability has by no means erased the remnant density ridges that had accumulated along corridors of intrametropolitan rail access. Nonetheless, physiographic variability has been shown to have a pervasive effect that is largely independent of distance-related factors. Even though the total effect of attractive landscape features and conditions is relatively subtle, it represents a solid substratum that selectively shapes the more intricate patterns of suburban population densities. The finding that physiographic factors play a measurable and qualitatively consistent role within given segments of Chicago's metropolitan region therefore confirms the central part of the study's original hypothesis. The importance of natural landscape features to suburban settlement is not an illusion. For some people a portion of the suburban dream has become reality, namely to live in surroundings that are physiographically attractive, if not unique. The presence of distinctive landscape elements reduces the banal concern of distance and attracts a greater number of residents than areas at similar distances that are physiographically less distinct and, therefore, less attractive.

Chapter 6

REGIONAL POPULATION DENSITIES AND PATTERNS OF CHANGE

To some degree, in most parts of America's inhabited domain the metropolis is almost everywhere.
—P.F. Lewis, "The Galactic Metropolis"

Regional Gradients

The analysis and results presented in the foregoing chapter have by and large justified the research design of breaking the metropolitan region into separate and, for analytical purposes, independent sectors. Little has been said about the relative size of the effect a given landscape term has vis-à-vis the general population distribution within the entire region. For the purpose of presenting a regional overview, sectoral boundaries are dropped and the entire suburban portion of the Chicago metropolitan area in northeastern Illinois is collapsed into the familiar straightjacket of the one-dimensional linear city.

Such regionwide analysis functions not only as a baseline for comparative metropolitan analyses, but also as a starting point for reviewing the geographic pattern of densities and the incidence of change across the suburban fringe of Chicago. This chapter then consists of two major parts: the first reviews the results of a set of linear regressions; the second explores through a series of maps the regional patterns of residential densities as well as the shifts that occurred between 1970 and 1980.

Regression Results and Interpretation

At the beginning of the decade in 1970 three of the five significant variables are physiographic variables (table 17). The first variables to enter are linear distance to Chicago's loop and commuter rail access. Lakes, rivers,

TABLE 17: Determinants of population density in Chicago's suburban region, 1970: coefficients and (standard errors) of landscape and access variables.

R^2	.499	.624	.640	.648	.651	.652	.654	.654	.655
Constant	9.472	7.581	7.634	7.486	7.411	7.217	7.151	7.181	7.190
Distance	**-.110**	**-.079**	**-.081**	**-.080**	**-.084**	**-.084**	**-.079**	**-.079**	**-.080**
	(.004)	(.004)	(.004)	(.004)	(.004)	(.004)	(.004)	(.004)	(.004)
Rail		**1.741**	**1.680**	**1.653**	**1.625**	**1.640**	**1.623**	**1.635**	**1.633**
		(.101)	(.099)	(.098)	(.098)	(.098)	(.099)	(.099)	(.099)
Lake			**.002**	**.003**	**.003**	**.003**	**.003**	**.003**	**.003**
			(.0003)	(.0004)	(.0004)	(.0004)	(.0004)	(.0004)	(.0004)
River				**.004**	**.004**	**.004**	**.004**	**.004**	**.004**
				(.001)	(.001)	(.001)	(.001)	(.001)	(.001)
Relief					**.003**	**.004**	**.004**	**.004**	**.004**
					(.001)	(.001)	(.001)	(.001)	(.001)
Highway						.190	.191	.180	.180
						(.093)	(.093)	(.094)	(.094)
Golf							*.001*	*.001*	*.001*
							(.001)	(.001)	(.001)
Forest								-.0003	-.0002
								(.0002)	(.0002)
Nature Park									-.00005
									(.0002)

.015 significant at ≥ 99% level of confidence
.015 significant at ≥ 95% level of confidence
.015 < 95% level of confidence

and relief variation are the other three significant variables. The effect of the four-mile-wide highway ribbon fails to sustain its initial level of significance. Except for distance, all other significant variables have a positive effect on estimated population densities. Ten years later there are a total six significant variables, with golf courses the only new variable to enter (table 18). Highway access remains altogether insignificant, as do forests and nature parks.

Of all physiographic landscape elements the presence of lakes exerts by far the most consistent influence throughout the region. In contrast, the relative importance of rivers has declined during the ten-year period. By 1980 it is surpassed by topographic variation; the initial coefficient of relief is twice the size when compared to 1970. The positive effect of golf courses on population densities is somewhat unexpected, since the sectoral analyses showed them to be a relatively inconspicuous factor. Overall the influence of physiographic variables has remained stable with occasional minor improvements, whereas the importance of commuter rail access has shrunk noticeably both in strength and size.

TABLE 18: Determinants of population density in Chicago's suburban region, 1980: coefficients and (standard errors) of landscape and access variables.

R^2	.509	.601	.624	.636	.642	.644	.646	.646	.646
Constant	9.725	8.118	8.179	8.021	7.899	7.810	7.665	7.661	7.665
Distance	-.109	-.083	-.086	-.092	-.090	-.088	-.086	-.086	-.086
	(.004)	(.004)	(.004)	(.004)	(.004)	(.004)	(.004)	(.004)	(.004)
Rail		1.478	1.406	1.350	1.331	1.307	1.319	1.317	1.317
		(.102)	(.0994)	(.098)	(.098)	(.098)	(.098)	(.099)	(.099)
Lake			.003	.003	.003	.003	.003	.003	.003
			(.0004)	(.0004)	(.0004)	(.0004)	(.0004)	(.0004)	(.0004)
Relief				.006	.005	.005	.005	.005	.005
				(.001)	(.001)	(.001)	(.001)	(.001)	(.001)
River					.004	.004	.004	.004	.004
					(.001)	(.001)	(.001)	(.001)	(.001)
Golf						.001	.001	.001	.001
						(.001)	(.001)	(.001)	(.001)
Highway							.141	.142	.143
							(.092)	(.093)	(.093)
Forest								.0003	.0005
								(.0002)	(.0002)
Nature Park									-.0003
									(.0002)

.015 significant at ≥ 99% level of confidence
.015 significant at ≥ 95% level of confidence
.015 < 95% level of confidence

A more detailed interpretation of the effect of each variable appears superfluous, since the preceding sectoral analyses have already dealt with the intraregional variability of factors. Even though the regional results should not directly be compared with the sectoral profiles, there are still a number of observations that dovetail with and indirectly strengthen some of the earlier conclusions. At the regional level as much as at the sectoral level the spatial pattern of suburban population densities is well accounted for by a combination of linear distance and intrametropolitan access, with commuter railroads being more important than highways. It is equally obvious that a number of physiographic elements are significant determinants, if modest in size, of residential densities. Their distribution may be greatly localized, as in the case of lakes; yet their effect is strong enough to be felt throughout the entire region. Based on the regional results, lakes, relief, rivers, and golf courses represent a set a physical landscape characteristics that do make a difference.

The foregoing analysis seems to converge on two principal factors that in recent urban studies have been dismissed too easily as determinants of present-day urban form and structure. It has been argued that differences in site quality other than mere distance and the sheer persistence of structural elements are important considerations in addition to growth and accessibility.[1] Site quality, as the first major characteristic of urban structure, encompasses a whole range of physiographic landscape attributes that distinguish urban locations. Much of the finer detail of physiographic diversity is cancelled out at the regional level, yet even at this scale a number of landscape variables remain significant. However, as the localized effects of lakes, rivers, relative topographic variation, and golf courses reverberate throughout the region, they do not form an undercurrent strong enough to upset the dominant effect of distance or access.

There is little doubt that commuter rail access in the Chicago area has had a cumulative effect on population densities and thus represents an element of structural persistence that is difficult to disregard. This effect is not evenly distributed across the region. As was detailed in the previous chapter, the density ridges along railroad corridors rose to different heights above the surrounding residential land. However, by the end of the decade the differences between the two have generally diminished; in some cases such as in the north and south sectors they have become entirely inconsequential.

The results thus far have shown that major groups of landscape features account for a significant portion of the spatial variation of residential densities in the Chicago suburban area. The question that remains is whether the distribution of landscape features is associated with any major changes in the pattern of suburban densities. The abstract estimation of linear density gradients has been central to the discussion of population densities of 1970 and 1980; yet it provides no clues as to the underlying factors of change. In order to draw some preliminary conclusions concerning the patterns of change during the 1970s, this method of analysis has been abandoned in favor of the more ordinary cartographic portrayal of population densities and change. Population change is the result of a series of compound events and isolated occurrences. Simple isoplethic maps will not disentangle the dense web of contributing factors and forces. At present, the modest objective of these maps is limited to identifying broad regional patterns and helping formulate initial hypotheses for future research.

Suburban Densities and Patterns of Change

The 1970 and 1980 distributions of population densities in the Chicago suburban area display distinct patterns that nonetheless have several structural features in common (figures 20 and 21). There are two sepa-

---

[1] B. Duncan, "Variables in Urban Morphology," in *Urban Sociology*, ed. E.W.Burgess and D.J. Bogue (Chicago: University of Chicago Press, 1967), ch. 1.

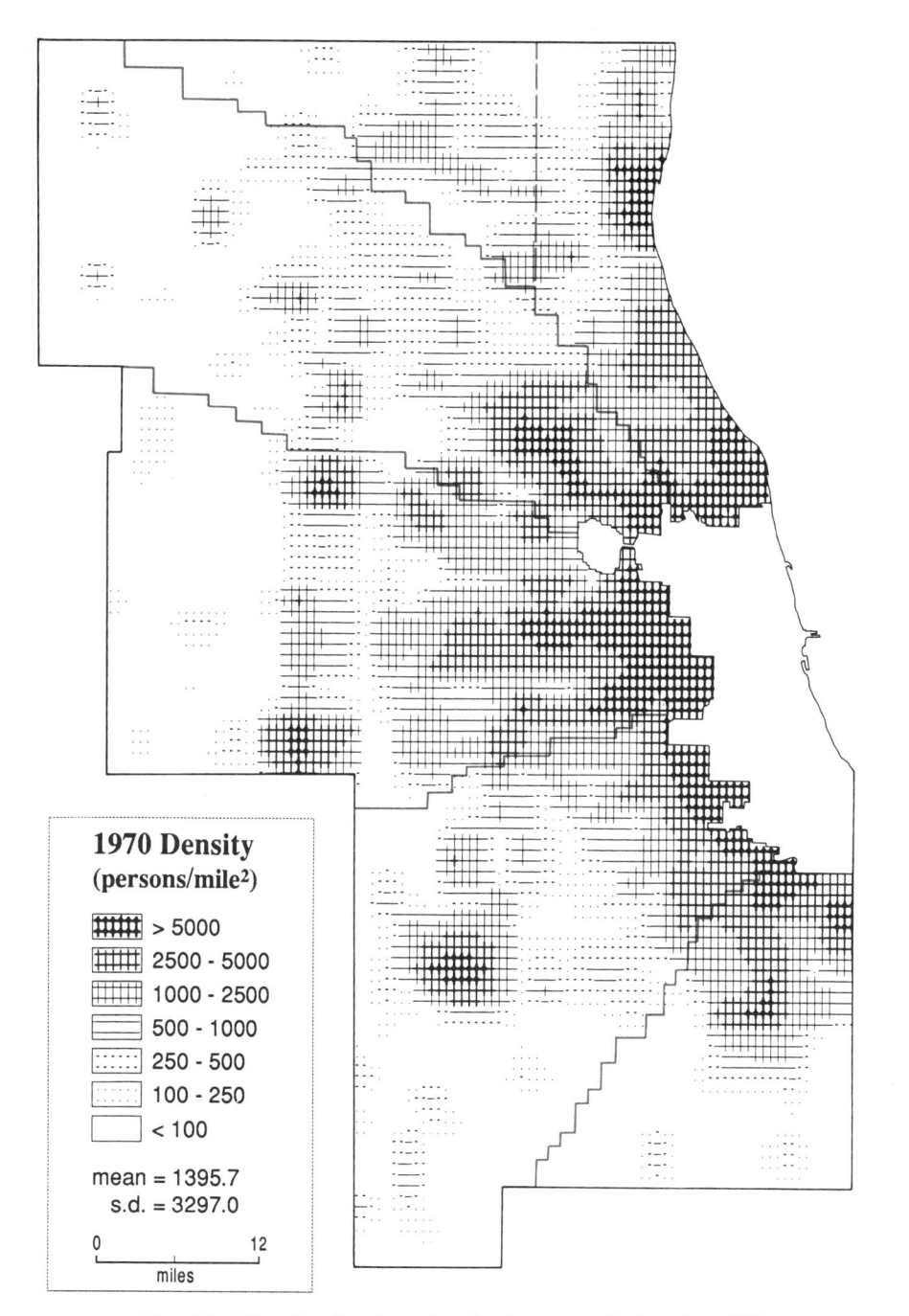

Fig. 20. *The distribution of suburban population in 1970.*

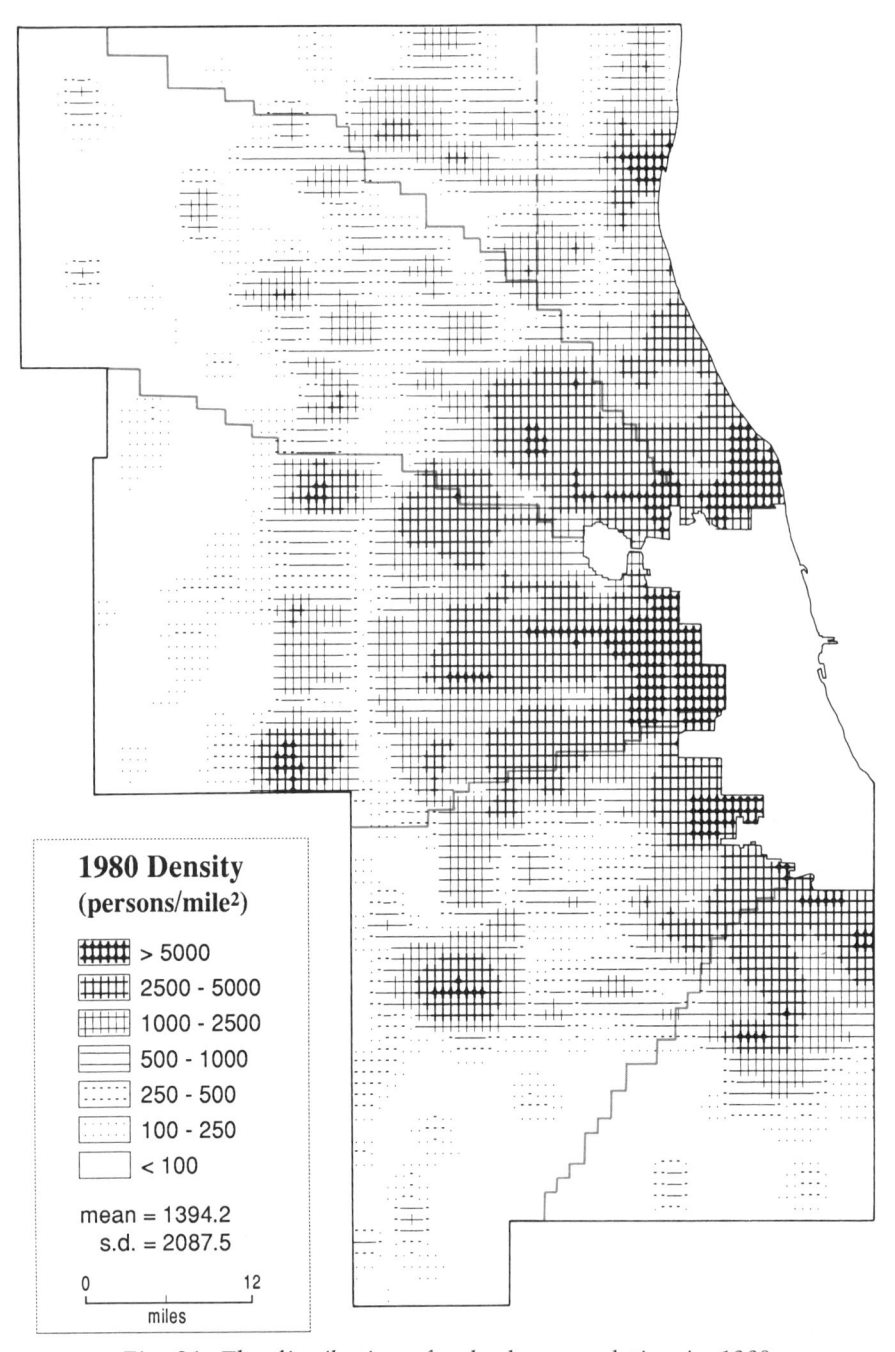

Fig. 21. *The distribution of suburban population in 1980.*

rate rings of residential densities. The first, inner ring of fairly high densities forms a contiguous area encircling Chicago. The second, outer ring of individual concentrations contains the string of settlements along the historic beltline of industrial satellite cities as well as the dispersed scatter of exurban communities in the region's northwestern portion. To the south of Chicago both belts lose their spatial separation and converge into an indistinguishable mix of suburban settlements and industrial satellites. The third distinguishing feature is the elevated density ridges that run across both rings into the regions perimeter, producing the characteristic radial lineaments along railroad corridors.

Without dwelling on the vagaries of cartographic data classification, a visual comparison of both maps still reveals noticeable shifts in the distribution of population that complement and reinforce previous interpretations and conclusions. By 1980 the radial density ridges are less prominent; at the same time there has been substantial infilling of interstitial areas at greater distances. As a result, the inner ring has increased its diameter, closing the gap between it and the outer ring. Residential densities rarely fall below 500 persons per square mile and are generally more evenly distributed without disturbing the general pattern of declining densities with increasing radial distance.

The growing equality between interstitial and corridor locations is not so much the result of densities in off-corridor areas catching up with on-corridor densities, as it is the result of falling densities in areas along transportation corridors. A larger discontinuity within the consolidating inner ring lies to the southwest of Chicago. As noted earlier, the pattern of residential expansion adheres to a transportation network whose geometry deviates substantially from the rest of the region. Instead of diverging along ordinary radial paths, two transportation corridors converge once more after leaving Chicago, creating a sort of residential vacuum with Chicago and Joliet as its end points.

As the inner ring expanded and lost some of its extreme internal density differentials, changes within the outer ring have been more restricted in both intensity and geographic reach. Residential expansion seems to be bound to existing settlement nuclei, be it the region's old factory towns such as Joliet, Aurora, and Elgin, or the post–World War II exurban refuges nestled within the remote lake districts to the north and northwest of Chicago.

The description of such general shifts in the regional patterns of population density needs to be substantiated by additional analysis. Toward this end, a look at the distribution of population gains and losses between 1970 and 1980 may answer the question whether there are any recognizable patterns of suburban population change. Since the number of four-section enumeration units is the same in both years, relative change in population and density are interchangeable terms. Population change is frequently expressed in relative terms. In this case, the geographic patterns of relative change appear quite random and provide no definite clues as to a set of common spatial attributes (figures 22 and 23). Measurements of relative population change remove, as much as cover up, considerable discrepancies in the size of enumeration districts and are insensitive to differences in starting population figures. It is not the size of areal units that is a major concern here; rather it is the gaping difference in population counts per unit. By employing relative change, an increase of 50 people in an area having 50 residents is given equal weight to the addition of 5,000 people in an area with an initial population of 5,000. Clearly, proportional differences are insufficient to describe the spatial distribution of change in a suburban setting.

Where the patterns of relative change remain inconclusive, the distribution of absolute population differences presents a more orderly regional pattern of change (figures 24 and 25). The most severe population losses have occurred within the older suburban areas close to Chicago and along several of the cardinal transportation axes, particularly to the northwest and west of the city. Secondary areas of decline can be found in the central portions of the old factory towns of Joliet, Aurora, Elgin, and Waukegan. Major population gains are concentrated in a broad band toward the perimeter of the inner suburban ring. There is no indication that growth has been channeled along the region's transportation corridors. Sizable increases are in areas away from the beaten path of commuter rail and multilane expressways. Within the outer ring one can observe two patterns. Population increases at the periphery of the industrial satellites are accompanied by pronounced losses at the town center. In contrast, many of the suburban villages and communities in the northwest and extreme northern sectors show an entirely different pattern of growth; instead of declining populations, central growth still outpaces growth at the periphery.

Generally, growth is greatest along the edges of the inner suburban ring. The profile of population growth is more subdued within the outer belt. Even in the booming lake district growth tends to be more evenly distributed so that differences between central and peripheral increases are less pronounced. The pattern within the inner belt shows steeper, more abrupt differentials. High levels of growth occur over large areas with precipitous declines in growth along their rims.

The distribution of suburban population change in the Chicago metropolitan area presents a pattern that contains features of standard suburban growth models mixed with irregular, less familiar elements. There is little indication that the overall pattern of change is informed by natural landscape features. Greatest loss of population occurs in both the older, close-in suburbs and the historic factory towns that ring the city of Chicago. Population gains are greatest at medium range. Thus the generalized, cross-

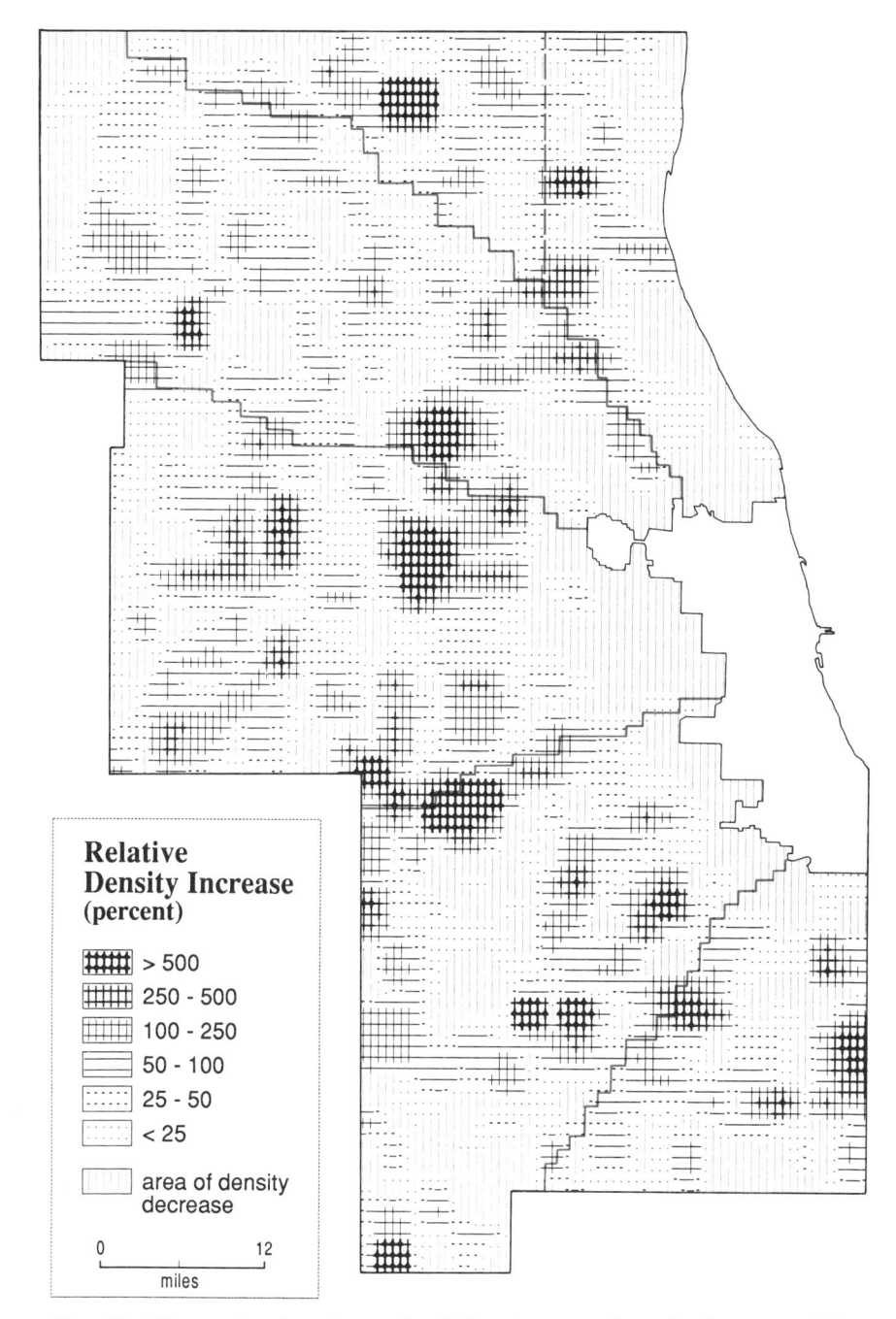

Fig. 22. *The regional pattern of relative increase in suburban population density between 1970 and 1980.*

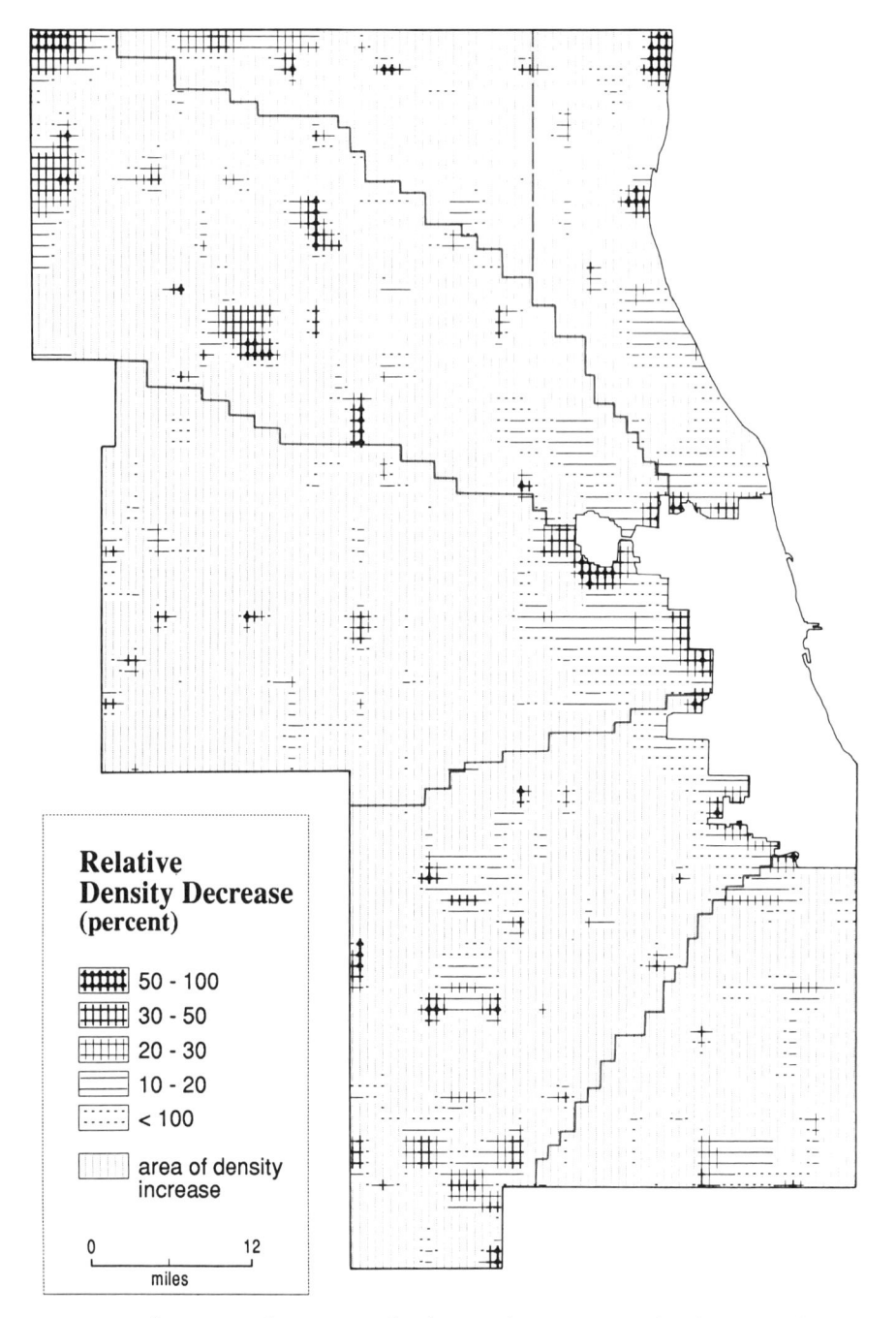

Fig. 23. The regional pattern of relative decrease in suburban population density between 1970 and 1980.

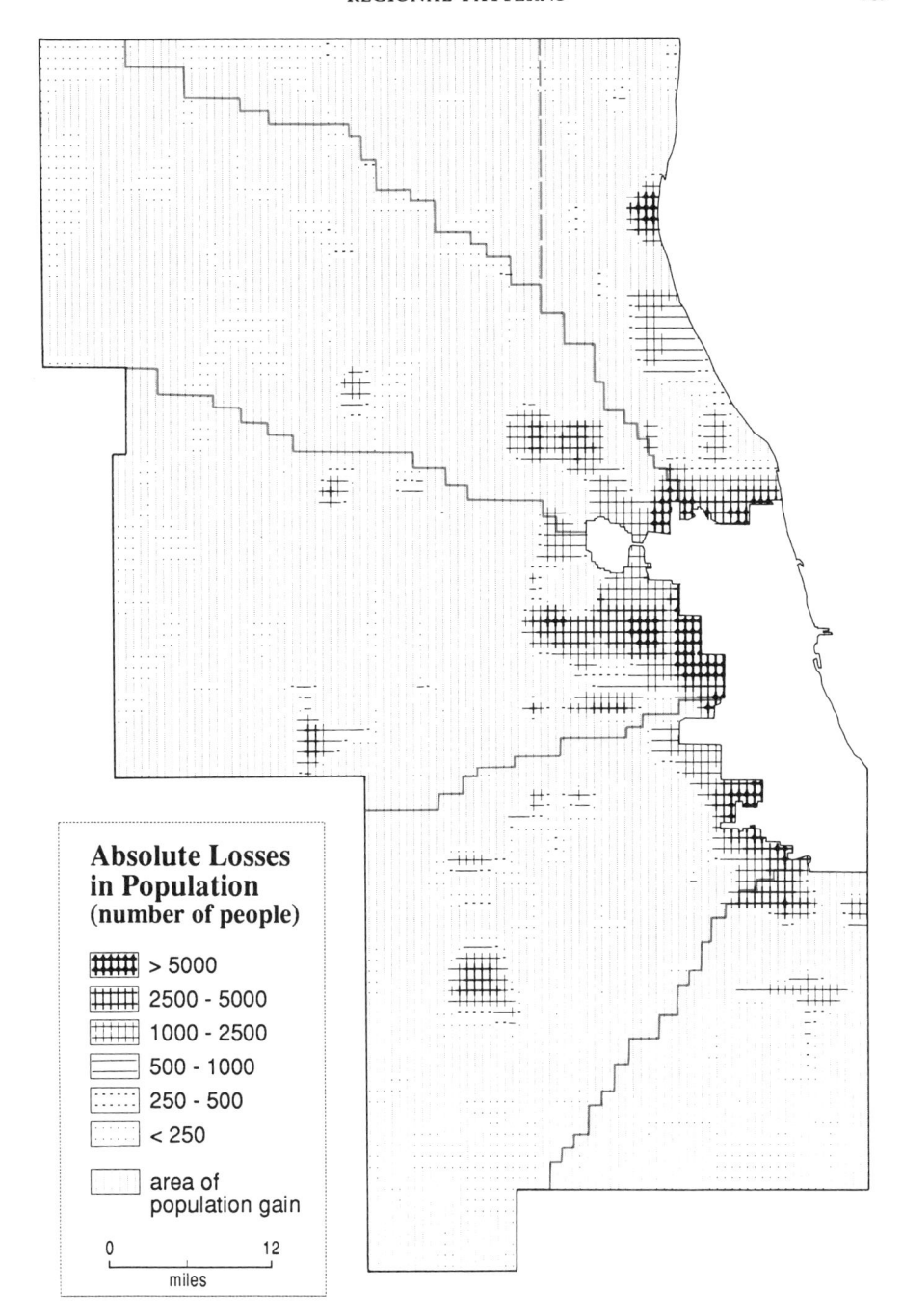

Fig. 24. The regional pattern of absolute losses in suburban population counts between 1970 and 1980.

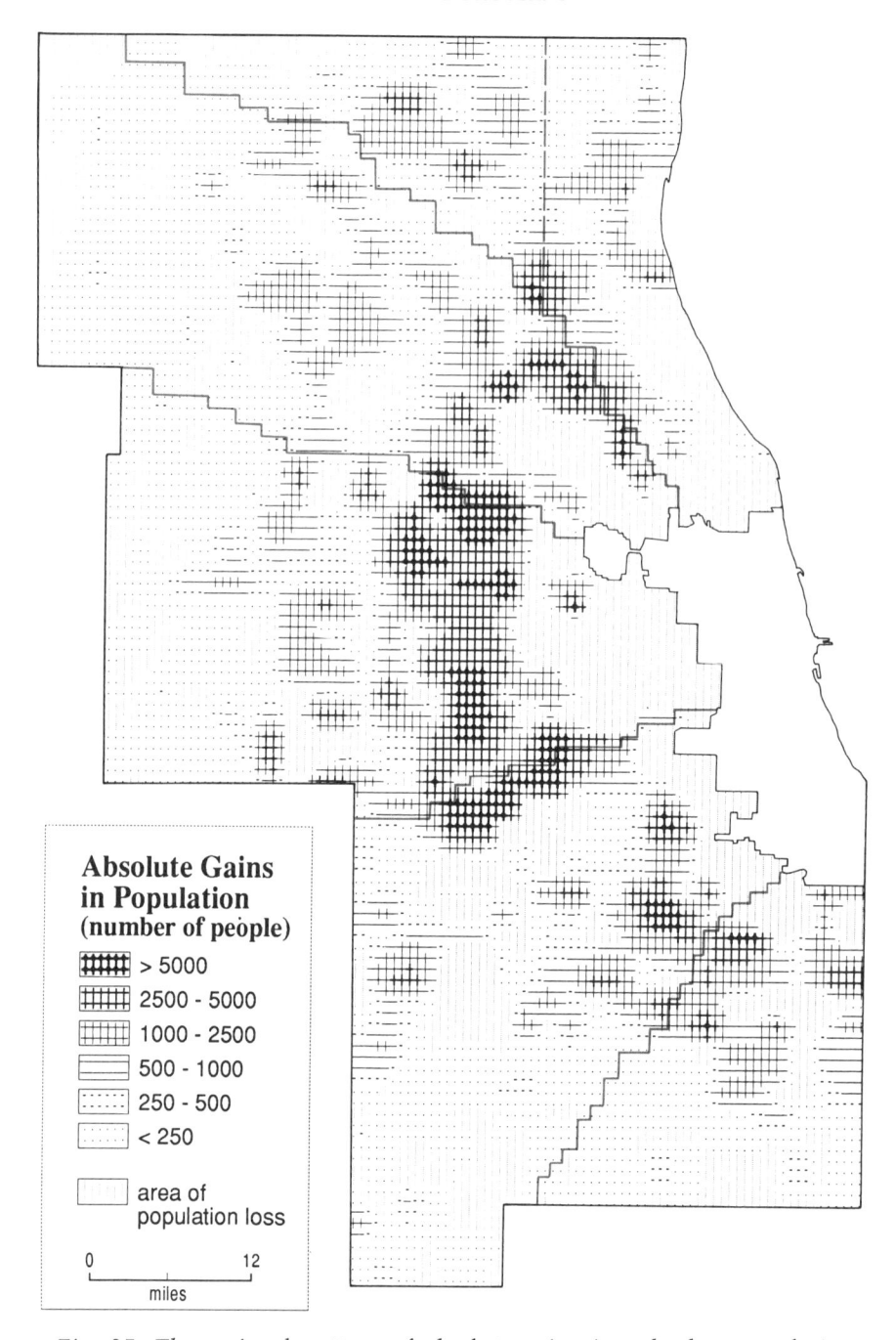

Fig. 25. The regional pattern of absolute gains in suburban population counts between 1970 and 1980.

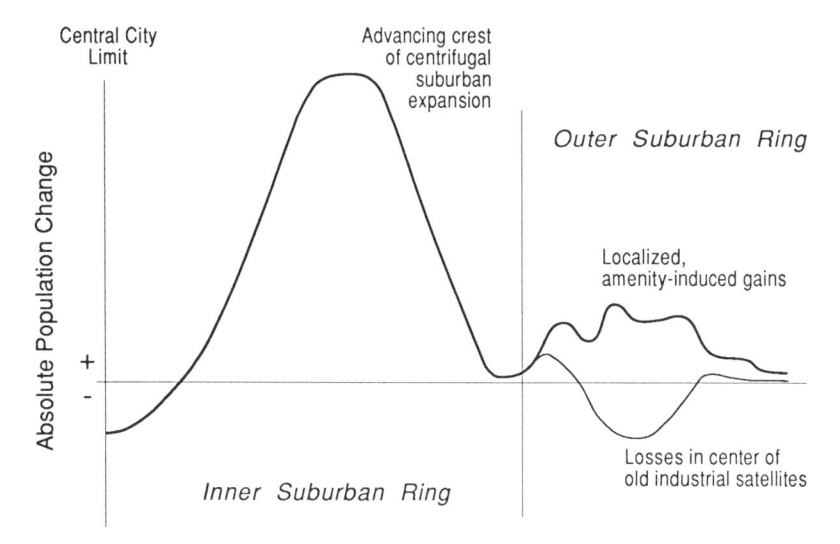

Fig. 26. Generalized, cross-sectional model of suburban population change in the Chicago metropolitan area, 1970–1980.

sectional profile of change follows a curvilinear, wavelike progression (figure 26). Similar patterns of suburban growth have been found elsewhere. For example the ecological perspectives of life cycle and peripheral expansion hold that distance from CBD and central city age are crucial factors in understanding the spatial variation in suburban growth and decline.[2] The concept of spread-backwash processes also relies on the wave analog model and sees the level of growth at the suburban periphery to be largely dependent on linear distance to the metropolitan core area.[3]

Toward the metropolitan periphery the growth curve splits along two paths. The first identifies the region's historical factory towns and is characterized by declining central populations and moderately growing outlying areas. The second path typifies the younger exurban communities whose fairly diffuse center of growth is surrounded by areas of modest increases. The latter growth profile may be taken as indirect evidence of environmental amenities inducing higher levels of population growth despite increased distance to the central city. Since this pattern repeats itself across the more distant, environmentally diversified northwestern reaches of the region, such a conclusion is not too farfetched. Yet, within the larger

---

[2] A.M. Guest, "Patterns of Suburban Population Growth, 1970-75," *Demography* 16 (1979): 401-415.

[3] S. Krakover, "Identification of Spatiotemporal Paths of Spread and Backwash," *Geographical Analysis* 15 (1983): 318-329.

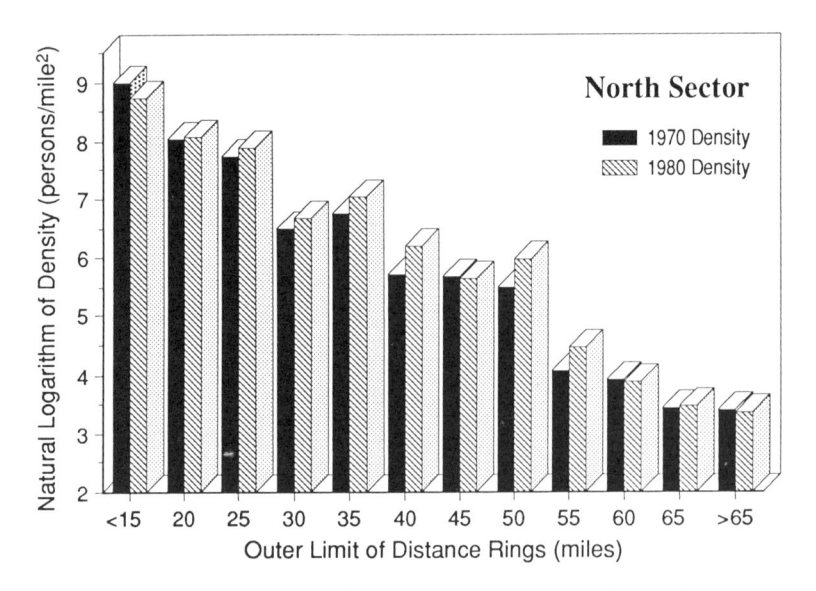

Fig. 27. Changing levels of suburban population density calculated as distance band averages in the north sector for 1970 and 1980.

regional context such spatial association must be considered an exception rather than the norm.

To summarize the shifts in density within the five sectors and the north-shore margin, each sector is divided into five-mile-wide rings or bands. Population densities are calculated as band averages. Two of the rings vary in width: the innermost ring represents areas within fifteen miles of the CBD; the outermost ring encompasses distances of greater than sixty-five miles. The principal features of the regional pattern of population change are easily recognized in each sector (figures 27-32). Without exception population density decline occurs within the interior distance bands. Even along the north-shore margin the rhythm of losses and gains conforms to the regional model of population change (figure 28a). The margin's particular status rests on its proximity to Lake Michigan. As noted in the previous chapter, the apparent decline of densities along a narrow coastal band may not be indicative of a diminished attraction to living near the lake (figure 28b). Since the decline occurs in an area that has a fairly long suburban history, it may indeed reflect the cyclical nature of population age and a residential housing market similar to that of the older suburban communities close to the city. Despite the relatively coarse zonal subdivision, each sectoral profile shows the telltale signs of the region's dual suburban character: the continued expansion of the interior ring of primary suburban settlement and the patchwork of peripheral suburban

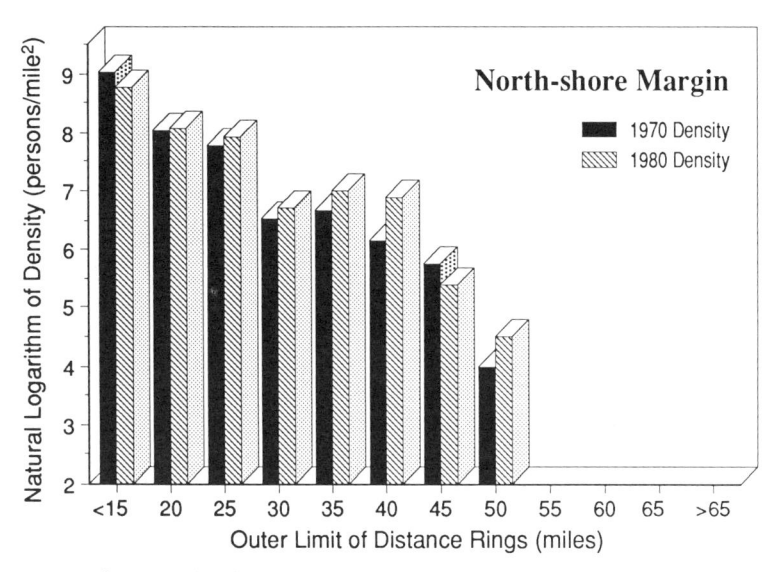

Fig. 28a. Changing levels of suburban population density calculated as distance band averages in the north-shore margin for 1970 and 1980.

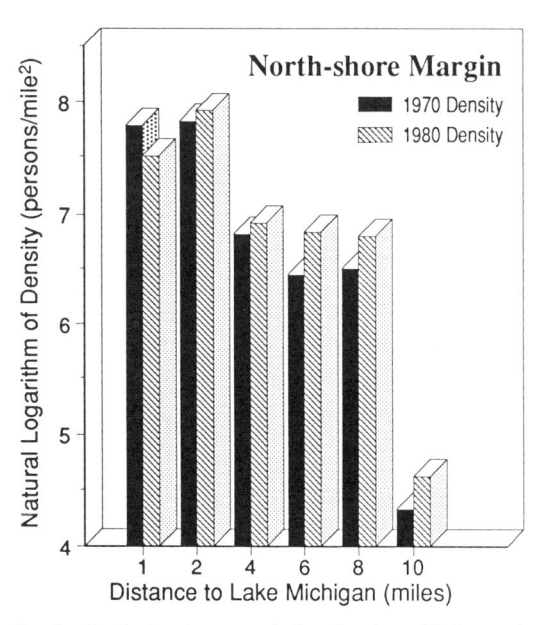

Fig. 28b. Changing levels of suburban population density with increasing distance from Lake Michigan within the north-shore margin for 1970 and 1980.

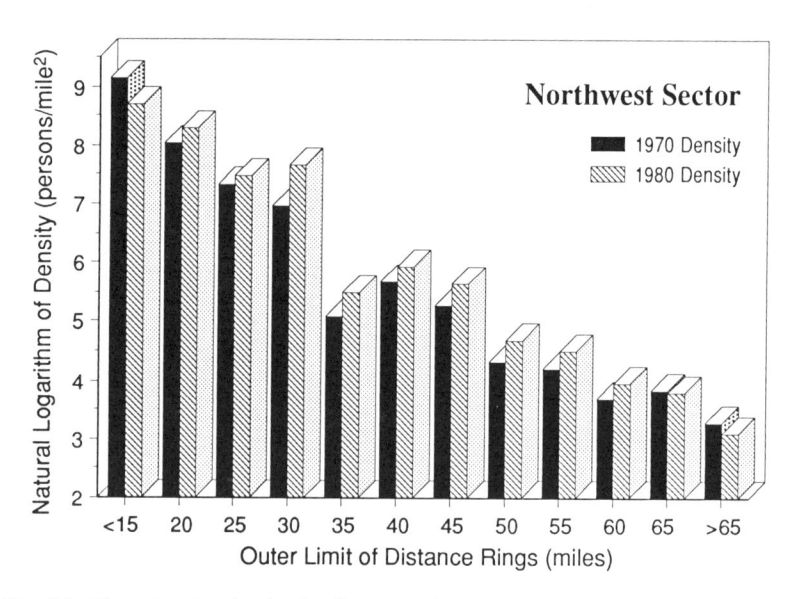

Fig. 29. Changing levels of suburban population density calculated as distance band averages in the northwest sector for 1970 and 1980.

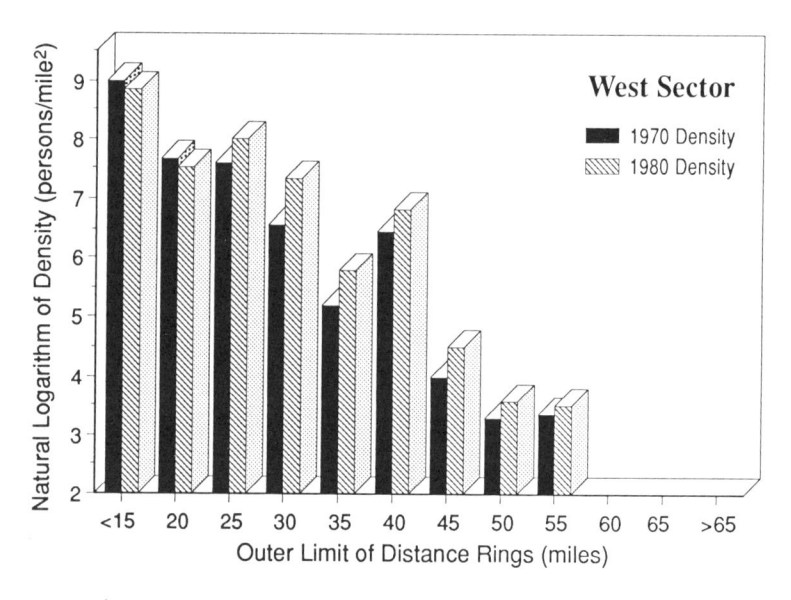

Fig. 30. Changing levels of suburban population density calculated as distance band averages in the west sector for 1970 and 1980.

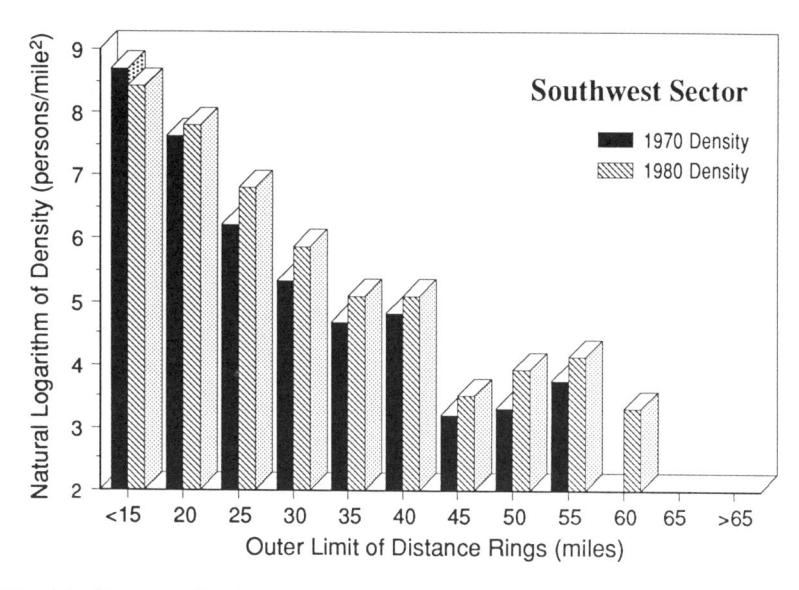

Fig. 31. Changing levels of suburban population density calculated as distance band averages in the southwest sector for 1970 and 1980.

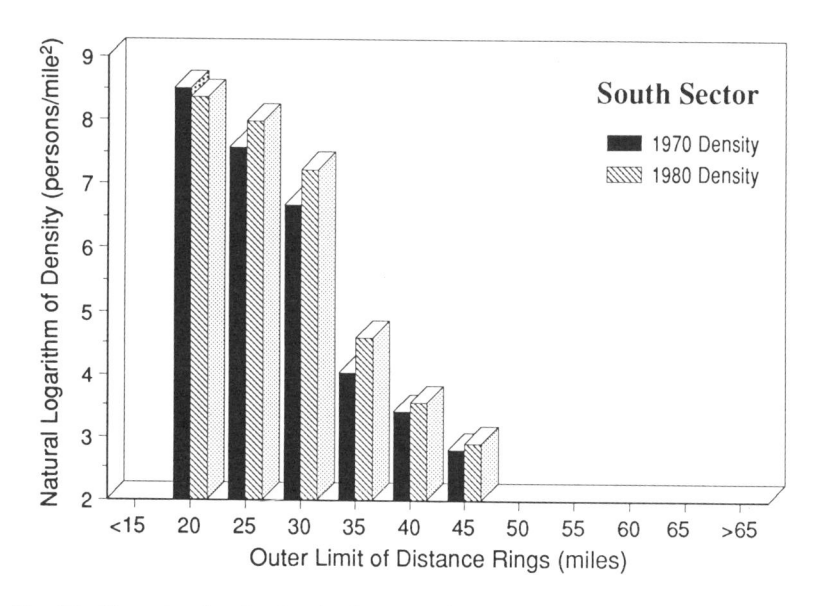

Fig. 32. Changing levels of suburban population density calculated as distance band averages in the south sector for 1970 and 1980.

growth that bypasses the old factory towns and reaches for the remote areas known for their attractive natural settings. These patterns of change complement previous research that established that amenity-rich areas within reach of metropolitan commuting grow faster than areas having no distinctive environmental amenities.[4] Yet, the patterns display more than the simple indiscriminate duality of growth defined by the absence or presence of environmental amenities.

This chapter's objective was not to prove or disprove that variations in landscape features and conditions are a major force in shaping the patterns of suburban population change. The cartographic evidence presented here shows that despite being the result of a complex set of processes, the spatial imprint of population growth and decline can be portrayed in rather straightforward terms. Next to distance from CBD there is clearly a host of other socioeconomic variables that will need to be considered in order to reach a more comprehensive and satisfactory explanation. As far as the role of natural landscape features is concerned, however, it is apparent that they function as a steady and silent companion keeping pace with suburban growth, but not actively directing it. For the most part the distribution of population growth and loss in Chicago's suburbs is independent of natural amenities. Only at the outer exurban periphery can landscape characteristics become a potent ingredient in the mixture of locational factors that encourage rapid residential development.

---

4   R. Lamb, *Metropolitan Impacts on Rural America*, University of Chicago Department of Geography Research Paper no. 162 (Chicago: University of Chicago Department of Geography, 1975), pp. 186-188.

Chapter 7

SUMMARY AND OUTLOOK

> *The process of urban growth is so complex as to defy total unraveling, regardless of the temporal or geographical setting, and because geographers and other scholars consequently are apt to select different analytical foci and to come up with a variety of interpretations.*
>
> —A. Pred, *Urban Growth and City-Systems*

This study's major objective was to explore the spatial association between selected natural landscape features and the pattern of suburban population densities across the Chicago metropolitan area in 1970 and 1980. The study's principal premise was that appealing natural landscape features have a significant effect on the distribution of suburban populations. Pronounced relief provides vistas; trees promise privacy; lakes and rivers offer serenity; and, like nature preservation areas, parks, and golf courses, support a variety of outdoor recreational activities. They are representative of landscape elements that appeal to suburban residents in general and, therefore, can be considered potential environmental amenities. As such they add another dimension to urban density models that most commonly use linear distance as the sole factor describing the behavior of density gradients.

The notion of living in suburban surroundings conjures up diverse images and expectations; yet near the center of the concept are low population densities and picturesque natural environs. "City-close and country-quiet" is more than a slogan advertising residential property outside the central city; it projects an image of residential life that sets the standard against which one holds alternative residential locations and circumstances. Differences between city and suburb are customarily couched in social, economic, and political terms. Studies have also focused on the assessment of environmental damage (e.g., pollution of air and water) and the accounting of their spillover effects on quality-of-life differentials among suburban and central-urban locations. The relatively permanent aspects of the natural environment, however, have rarely been viewed as useful diagnostic variables that could provide additional insights into the spatial variation of suburban populations and patterns of growth.

The analyses of 1970 and 1980 data on the distribution of population in the Chicago suburban area demonstrated that physiographic landscape elements have a statistically measurable influence on the spatial variation of population densities. A sequence of sectoral cross sections was performed, followed by a regional summary and an exploratory cartographic examination of observable patterns of suburban density change. The sectoral division was largely based on an intraurban accessibility criterion that determined commuter railroad lines, as a historically persistent factor of urban structure, to be associated with abnormally high density levels. This division also made it possible to account for the intraregional imbalance in the distribution of landscape features selected for this study.

The specific effects, whether in size, direction, or duration, of the physiographic features on population density were far from uniform across the region. Whereas fluctuations in size were not unexpected, there were notable differences in the direction and duration of the effect. In most cases the effect was positive. Increased levels of physiographic features were accompanied by increases in density above the level determined by any of the three intraurban accessibility measures of linear distance, suburban commuter railroad access, and highway ribbons. Forests constitute the only landscape variable that had a negative influence on estimated population densities. Significantly, the direction of influence of a given landscape variable did not change through time or from sector to sector. Finally and reassuringly, none of the selected landscape variables remained altogether insignificant.

This study identified three principal types of landscape variables. The types are differentiated by their characteristic differences in timing and duration vis-à-vis metropolitan growth and development. The threefold typology consists of:

1. Landscape fixtures. Their effect is established during the early stages of residential expansion along the suburban periphery and remains fairly constant throughout the next stages of suburban growth. A landscape fixture is associated with leap-frogging, discontinuous residential growth seeking to occupy locations previously reserved for exurban resorts and secondary homes nestled in the remote corners of a metropolitan region.
2. Transitional features. Their effect is strongest during the early stages of suburban growth. As growth continues their intrinsic value diminishes, owing to crowding and associated damages.

3. Emergent features. Previously bypassed, these features assume a growing role during the later stages of suburban development and expansion.

In the case of suburban Chicago, lakes and rivers represent the first type. Forests, parks, and nature preservation areas characterize the second type. Relief and golf courses belong to the last group. This typology must be considered largely preliminary and tentative and cannot be expected to apply elsewhere. However, it provides a conceptual baseline for analyzing the patterns of contemporary growth and development of other metropolitan areas in the United States.

The regional summary showed that the effects of several landscape features are sufficiently strong and pervasive so as to account for a significant portion of the spatial variation in population density of the entire region. It indirectly confirmed the usefulness of viewing natural landscape components as fixtures and as emergent or transitional features. None of the features classified as transitional in the sectoral analyses played a measurable role in the regional summary. Even though much of the physiographic detail is lost at the regional scale, it is clear that Chicago's suburban population has undergone extensive locational sorting and redistribution. The historic density pattern that extended along the commuter rail corridors has lost much of its definition. Density differentials between areas within railroad corridors and those outside decreased markedly. As the density ridges accumulated during previous phases of suburban growth and expansion eroded, natural landscape features did not become more significant as locational attributes.

The patterns of absolute population gains and losses underscored this conclusion. The major landscape components considered here were not a major underlying component of population/density change. The patterns of change revealed two rings of suburban development and a gradually shrinking, still sparsely populated perimeter. Within the inner ring at close range to Chicago, population losses were far more common than gains. Increases in population were predominantly clustered near the inner ring's periphery. The pattern of change within the outer suburban ring displayed two variations. The first displayed a pattern of cratering central densities and expanding peripheries, a familiar pattern found in regions with an aging metropolis. This distribution of change was mostly confined to Chicago's industrial satellites. The second pattern occurred in areas of superior natural endowment, such as in the northwest, and was characterized by continued, somewhat diffuse central growth accompanied by only marginally lower increases at the edges. No doubt these patterns, though cartographically verified, need further inspection and more detailed analysis. As it stands, they confirm previous research that concluded that the direction and size of growth in areas within reach of metropolitan commuting will largely be determined by the area's level of natural amenity endowment rather than such conventional inputs to production as resources and labor.

Of the many issues that future research could take up in order to complement and amend this study, only a few shall be discussed here. There are three main tracks such research can follow. The first would increase the time depth of the Chicago case study; different time periods during Chicago's explosive growth should be studied for comparative purposes. Given the cyclical nature of urban expansion, it seems essential to examine the social, economic, and political circumstances under which landscape amenities emerge as crucial determinants of residential location. Such studies would build the longitudinal record of urban growth and expansion that is crucial to determining whether environmental amenities have indeed been important locational factors in the formation of America's suburban landscape.

This course of inquiry should be accompanied by studies that compile detailed local accounts and case histories of suburban communities. Such investigations must tackle the question of whether there are patterns of socioeconomic stratification underlying the sequence of residential expansion into amenity-rich areas. Furthermore, examination of communities at various distances from the central city will be able to detect temporal and locational variations in the perceived significance and actual value of comparable environmental assets. For example, what distinguishes the amenity-related experiences of communities such as Riverside, Frankfort, or Holiday Hills?

The second track would investigate whether similar spatial and temporal patterns exist in other metropolitan areas. Such comparisons may lead to a reinterpretation or replacement of the standard four-stage model of metropolitan growth based largely on transportation technology. If proclaimed cultural dispositions are indeed valid constructs, did they assert themselves in previous developmental periods? If so, in what form and to what extent?

The third track would directly deal with the life span of specific environmental amenities. Is there a cycle of utility during which the value of a landscape component rises, declines, ceases to exist, or reemerges? Under what circumstances can an amenity level be maintained and protected? The trade-off between density and preservation of an agreed-upon amenity value takes place in a political, social, and economic context and is secured through such measures as zoning, taxation, and so forth, or by renewed relocation. As this study showed, the effect of proximity to Lake Michigan, so obvious at the beginning of the decade, had diminished considerably by

1980. At what point does population density become an unreliable indicator of environmental amenities and their inferred value; and at what point does increasing density reduce an amenity's value?

Future studies will have to develop a broadened perspective that can deal with, and possibly dissolve and supersede, conflicting research philosophies and methodologies. There is a general need for a more flexible approach that allows one to explore the overlapping realms of social, economic, and political structures and agents and the roles they play in the valuation of natural landscapes within metropolitan settings during different periods of growth and expansion. Renewed attempts should be made at joining on a conceptual level local, regional, and national scales of inquiry. There is a need, both theoretical and practical, for a coherent and unifying statement of urban growth and development that can deal both with issues of national urban system organization and with the structure of individual cities and metropolitan regions. At the same time, this reorientation of urban research needs to address the apparent duality of the American urban experience, which consists of a strong central European tradition of city life as well as the American preference for a suburban way of life.

Finally, there remains the issue of methodology. This study provides an indirect test of the principles and standards of a geographic information system (GIS) even though the construction of a GIS and its implementation did not constitute the study's primary objectives. Data gathered from diverse sources and at different scales were assigned to a fairly uniform spatial matrix, which allowed the analysis of attributional characteristics of each cell as well as a comparison between cells. Instead of imposing an abstract, geometrically fixed spatial grid as is customary in GIS applications, the existing, slightly irregular layout of the public land survey was used to determine the region's spatial division. What the study may have lost from the survey's evident spatial biases and distortions, it more than gained in functional fidelity. A comparison between the two approaches seems inevitable. Such effort would also establish strategies useful in tackling the threefold research agenda outlined above.

Appendix A

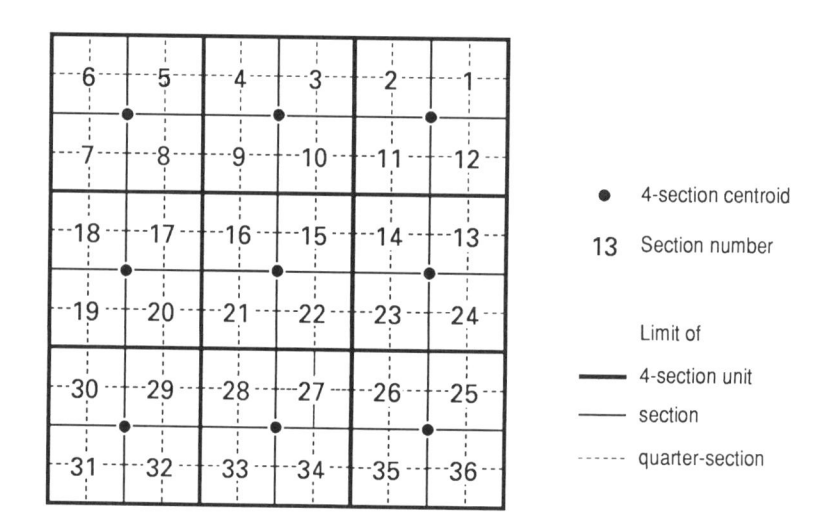

6	5	4	3	2	1
7	8	9	10	11	12
18	17	16	15	14	13
19	20	21	22	23	24
30	29	28	27	26	25
31	32	33	34	35	36

- ● 4-section centroid
- 13 Section number

Limit of
- ——— 4-section unit
- ——— section
- - - - - quarter-section

Pattern of Spatial Aggregation within a Township.

Appendix B

Descriptive Statistics of Each Variable within the Five Sectors, the North-Shore Margin, and the Suburban Region.

North Sector:

Variable	N	mean	s.d.	sum	minimum	maximum
Density 1970	159	6.1259	1.8913	974.0325	2.6912	10.8932
Density 1980	160	6.2728	1.8235	1003.6522	2.7944	9.5612
Distance	160	37.2900	13.5200	5967.6100	8.8100	65.2200
Rail	160	0.6100	0.4600	99.0000	0.0000	1.0000
Highway	160	0.3500	0.4800	57.0000	0.0000	1.0000
Forest	160	276.1800	200.2300	44190.0000	0.0000	970.0000
Lake	160	93.2500	211.2200	14920.0000	0.0000	1270.0000
Nature Park	160	87.0000	190.8200	13920.0000	0.0000	1080.0000
Golf	160	4.0300	8.1700	645.0000	0.0000	40.0000
River	160	24.4600	36.5400	3905.0000	0.0000	147.0000
Relief	160	70.7500	37.5600	11320.0000	0.0000	210.0000

North-Shore Margin:

Variable	N	mean	s.d.	sum	minimum	maximum
Density 1970	92	7.0236	1.6252	646.1768	3.1069	10.8932
Density 1980	93	7.0692	1.5456	657.4380	2.7944	9.5612
Distance to CBD	93	29.1466	10.8598	2710.6397	8.8104	46.7418
Distance to Lake	93	4.4731	2.7842	416.0000	1.0000	10.0000
Rail	93	0.7473	0.4274	69.5000	0.0000	1.0000
Highway	93	0.6129	0.4897	57.0000	0.0000	1.0000
Forest	93	250.8602	198.3445	23330.0000	0.0000	970.0000
Lake	93	41.0752	88.4831	3820.0000	0.0000	490.0000
Nature Park	93	118.3870	204.5902	11010.0000	0.0000	1060.0000
Golf	93	6.4946	9.9188	604.0000	0.0000	40.0000
River	93	28.3870	39.8658	2640.0000	0.0000	147.0000
Relief	93	58.1720	30.4297	5410.0000	0.0000	120.0000

Northwest Sector:

Variable	N	mean	s.d.	sum	minimum	maximum
Density 1970	224	5.1300	1.8344	1149.1366	2.5649	9.8430
Density 1980	225	5.3789	1.9021	1210.2589	2.0368	9.0302
Distance	227	44.0838	14.0583	10007.0364	12.2875	69.8913
Rail	227	0.3766	0.4682	85.5000	0.0000	1.0000
Highway	227	0.2731	0.4465	62.0000	0.0000	1.0000
Forest	227	336.5638	258.5042	76400.0000	0.0000	1260.0000
Lake	227	27.8414	56.3350	6320.0000	0.0000	310.0000
Natute Park	227	85.5506	274.9883	19420.0000	0.0000	1700.0000
Golf	227	2.2158	6.6126	503.0000	0.0000	53.0000
River	227	20.7004	33.4900	4699.0000	0.0000	185.0000
Relief	227	84.8458	51.2960	19260.0000	10.0000	250.0000

West Sector:

Variable	N	mean	s.d.	sum	minimum	maximum
Density 1970	241	5.7395	2.1579	1383.2238	2.0862	9.9738
Density 1980	242	6.0756	2.0798	1470.3183	1.2027	9.5704
Distance	242	33.3424	12.0709	8068.8721	6.6513	52.8371
Rail	242	0.4669	0.4809	113.0000	0.0000	1.0000
Highway	242	0.4297	0.4960	104.0000	0.0000	1.0000
Forest	242	245.9504	231.9378	59520.0000	0.0000	1160.0000
Lake	242	4.9586	11.0917	1200.0000	0.0000	80.0000
Nature Park	242	79.5041	179.5122	19240.0000	0.0000	970.0000
Golf	242	3.3595	8.0572	813.0000	0.0000	49.0000
River	242	24.9338	40.9374	6034.0000	0.0000	202.0000
Relief	242	64.3388	33.8255	15570.0000	10.0000	180.0000

Southwest Sector:

Variable	N	mean	s.d.	sum	minimum	maximum
Density 1970	213	5.3188	2.0427	1132.9113	1.2653	10.6798
Density 1980	218	5.5847	2.0242	1217.4776	1.6707	9.1803
Distance	218	32.1574	11.2021	7010.3191	8.5123	55.9244
Rail	218	0.3692	0.4630	80.5000	0.0000	1.0000
Highway	218	0.4724	0.5003	103.0000	0.0000	1.0000
Forest	218	156.2385	169.8022	34060.0000	0.0000	830.0000
Lake	218	20.0917	64.5294	4380.0000	0.0000	600.0000
Nature Park	218	125.0000	299.3483	27250.0000	0.0000	2110.0000
Golf	218	1.8073	5.3666	394.0000	0.0000	30.0000
River	218	43.4266	55.0274	9467.0000	0.0000	260.0000
Relief	218	68.1192	38.0352	14850.0000	0.0000	180.0000

South Sector:

Variable	N	mean	s.d.	sum	minimum	maximum
Density 1970	97	5.1494	2.2422	499.4928	2.6039	8.9710
Density 1980	97	5.4535	2.1753	528.9947	2.4426	8.9062
Distance	97	31.1386	7.1603	3020.4482	15.5474	43.6454
Rail	97	0.2989	0.4487	29.0000	0.0000	1.0000
Highway	97	0.5876	0.4948	57.0000	0.0000	1.0000
Forest	97	200.2061	187.2385	19420.0000	0.0000	880.0000
Lake	97	5.7731	11.5321	560.0000	0.0000	50.0000
Nature Park	97	118.7628	303.8614	11520.0000	0.0000	1520.0000
Golf	97	3.2371	8.1250	314.0000	0.0000	34.0000
River	97	17.4020	29.0666	1688.0000	0.0000	165.0000
Relief	97	49.7938	20.2061	4830.0000	0.0000	100.0000

Suburban Region:

Variable	N	mean	s.d.	sum	minimum	maximum
Density 1970	903	5.5009	2.0450	4967.3426	1.2653	10.8932
Density 1980	911	5.7651	2.0121	5252.0288	1.2027	9.5704
Distance	913	36.2853	13.1546	33128.5680	6.6513	69.8913
Rail	913	0.4348	0.4768	397.0000	0.0000	1.0000
Highway	913	0.3778	0.4851	345.0000	0.0000	1.0000
Forest	913	257.6122	229.8077	235200.0000	0.0000	1260.0000
Lake	913	30.4162	102.9927	27770.0000	0.0000	1270.0000
Nature Park	913	100.4271	258.0020	91690.0000	0.0000	2110.0000
Golf	913	29.5509	73.7982	26980.0000	0.0000	530.0000
River	913	28.2486	43.6591	25791.0000	0.0000	260.0000
Relief	913	70.0657	40.7387	63970.0000	0.0000	250.0000

APPENDIX B: DESCRIPTIVE STATISTICS

North Sector: Pearson Correlation Coefficients / PROB > |R| under H0: RHO=0 / N= Number of Observations

	D'70	D'80	Distance	Rail	H'way	Forest	Lake	N'Park	Golf	River	Relief
Density 1970	1.00000										
Density 1980	0.92556 0.0001 159	1.00000									
Distance	-0.75351 0.0001 159	-0.73259 0.0001 160	1.00000								
Rail	0.50581 0.0001 159	0.45523 0.0001 160	-0.49711 0.0001 160	1.00000							
Highway	0.21571 0.0063 159	0.24997 0.0014 160	-0.52961 0.0001 160	0.11937 0.1327 160	1.00000						
Forest	-0.15976 0.0443 159	-0.02815 0.7238 160	0.07022 0.3776 160	0.13823 0.0813 160	0.07697 0.3333 160	1.00000					
Lake	0.12886 0.1055 159	0.17183 0.0298 160	0.16588 0.0361 160	0.03421 0.6676 160	-0.25693 0.001 160	0.30821 0.0001 160	1.00000				
Nature Park	0.06192 0.4381 159	0.09207 0.2469 160	-0.13907 0.0795 160	0.01151 0.8852 160	0.27313 0.0005 160	0.28304 0.0003 160	0.02111 0.7911 160	1.00000			
Golf	0.33075 0.0001 159	0.35077 0.0001 160	-0.38662 0.0001 160	0.24353 0.0019 160	0.36067 0.0001 160	0.02428 0.7606 160	-0.07192 0.3661 160	0.1614 0.0415 160	1.00000		
River	-0.06705 0.4011 159	0.01172 0.883 160	-0.07943 0.3181 160	-0.00861 0.9139 160	0.26184 0.0008 160	0.32655 0.0001 160	-0.12832 0.1059 160	0.36349 0.0001 160	-0.00682 0.9318 160	1.00000	
Relief	-0.42144 0.0001 159	-0.33615 0.0001 160	0.58923 0.0001 160	-0.22343 0.0045 160	-0.30769 0.0001 160	0.08802 0.2684 160	0.03608 0.6506 160	-0.05488 0.4907 160	-0.16004 0.0432 160	0.07763 0.3292 160	1.00000

APPENDIX B: DESCRIPTIVE STATISTICS

North-Shore Margin: Pearson Correlation Coefficients / PROB > |R| under H0: RHO=0 / N= Number of Observations

	D'70	D'80	Distance	to Lake	Rail	H'way	Forest	Lake	N'Park	Golf	River	Relief
Density 1970	1.00000											
Density 1980	0.87501 0.0001 92	1.00000										
Distance	-0.7043 0.0001 92	-0.71279 0.0001 93	1.00000									
Dist. to Lake	-0.51537 0.0001 92	-0.35015 0.0006 93	0.16712 0.1094 93	1.00000								
Rail	0.65412 0.0001 92	0.51685 0.0001 93	-0.53325 0.0001 93	-0.50121 0.0001 93	1.00000							
Highway	-0.26624 0.0103 92	-0.13906 0.1837 93	-0.16889 0.1056 93	0.4387 0.0001 93	-0.16079 0.1236 93	1.00000						
Forest	-0.3031 0.0033 92	-0.1207 0.2491 93	0.02606 0.8042 93	0.0896 0.3931 93	0.02246 0.8308 93	0.33358 0.0011 93	1.00000					
Lake	0.12386 0.2395 92	0.02284 0.828 93	0.08566 0.4142 93	-0.28358 0.0059 93	0.1725 0.0982 93	-0.29381 0.0043 93	0.08535 0.4159 93	1.00000				
Nature Park	-0.06505 0.5378 92	0.00963 0.927 93	-0.02245 0.8309 93	0.1105 0.2917 93	-0.08239 0.4324 93	0.25841 0.0124 93	0.34922 0.0006 93	-0.01539 0.8836 93	1.00000			
Golf	0.20743 0.0473 92	0.25069 0.0154 93	-0.2703 0.0088 93	-0.12035 0.2505 93	0.19386 0.0626 93	0.22557 0.0297 93	0.11492 0.2727 93	-0.06439 0.5397 93	0.12006 0.2517 93	1.00000		
River	-0.15325 0.1447 92	-0.04122 0.6948 93	-0.06899 0.5111 93	0.29515 0.0041 93	-0.21042 0.0429 93	0.29504 0.0041 93	0.37125 0.0002 93	-0.13194 0.2074 93	0.48912 0.0001 93	-0.05464 0.6029 93	1.00000	
Relief	-0.35273 0.0006 92	-0.176 0.0915 93	0.53086 0.0001 93	0.11039 0.2922 93	-0.29911 0.0036 93	-0.12094 0.2482 93	0.10003 0.3401 93	0.32531 0.0015 93	-0.01445 0.8907 93	-0.05135 0.625 93	-0.07011 0.5043 93	1.00000

APPENDIX B: DESCRIPTIVE STATISTICS

Northwest Sector: Pearson Correlation Coefficients / PROB > |R| under HO: RHO=0 / N= Number of Observations

	D'70	D'80	Distance	Rail	H'way	Forest	Lake	N'Park	Golf	River	Relief
Density 1970	1.00000										
Density 1980	0.95049 0.0001 224	1.00000									
Distance	-0.7883 0.0001 224	-0.79542 0.0001 225	1.00000								
Rail	0.4648 0.0001 224	0.44977 0.0001 225	-0.27889 0.0001 227	1.00000							
Highway	0.33473 0.0001 224	0.34427 0.0001 225	-0.51432 0.0001 227	0.02429 0.7159 227	1.00000						
Forest	-0.00447 0.947 224	0.00332 0.9605 225	0.02321 0.728 227	0.01531 0.8186 227	-0.24827 0.0002 227	1.00000					
Lake	0.20163 0.0024 224	0.22459 0.0007 225	-0.11552 0.0824 227	-0.01433 0.83 227	-0.1717 0.0095 227	0.31904 0.0001 227	1.00000				
Nature Park	0.15489 0.0204 224	0.17632 0.008 225	-0.31051 0.0001 227	-0.00823 0.9018 227	0.21389 0.0012 227	0.18194 0.006 227	0.02945 0.6589 227	1.00000			
Golf	0.22674 0.0006 224	0.23867 0.0003 225	-0.17221 0.0093 227	0.05866 0.3791 227	0.0279 0.6759 227	-0.06904 0.3004 227	0.0705 0.2902 227	-0.07113 0.2859 227	1.00000		
River	0.22589 0.0007 224	0.23591 0.0004 225	-0.14144 0.0332 227	0.05802 0.3843 227	0.01881 0.7781 227	0.29188 0.0001 227	0.0175 0.7931 227	0.11491 0.0841 227	-0.02186 0.7432 227	1.00000	
Relief	-0.16617 0.0128 224	-0.10716 0.1089 225	0.29044 0.0001 227	-0.09752 0.143 227	-0.33813 0.0001 227	0.24639 0.0002 227	0.03441 0.606 227	-0.06365 0.3398 227	-0.01771 0.7908 227	0.06004 0.3679 227	1.00000

APPENDIX B: DESCRIPTIVE STATISTICS

West Sector: Pearson Correlation Coefficients / PROB > |R| under HO: RHO=0 / N= Number of Observations

	D'70	D'80	Distance	Rail	H'way	Forest	Lake	N'Park	Golf	River	Relief
Density 1970	1.00000										
Density 1980	0.94905 0.0001 241	1.00000									
Distance	-0.82957 0.0001 241	-0.80842 0.0001 242	1.00000								
Rail	0.82519 0.0001 241	0.79686 0.0001 242	-0.77728 0.0001 242	1.00000							
Highway	0.30894 0.0001 241	0.25218 0.0001 242	-0.36418 0.0001 242	0.22504 0.0004 242	1.00000						
Forest	0.17529 0.0064 241	0.18609 0.0037 242	-0.09276 0.1503 242	0.14574 0.0234 242	-0.11897 0.0646 242	1.00000					
Lake	0.10132 0.1167 241	0.17709 0.0057 242	-0.1392 0.0304 242	0.19811 0.002 242	0.03341 0.6051 242	0.10042 0.1192 242	1.00000				
Nature Park	0.24184 0.0001 241	0.24931 0.0001 242	-0.2786 0.0001 242	0.23797 0.002 242	0.08954 0.165 242	0.31638 0.0001 242	0.29966 0.0001 242	1.00000			
Golf	0.24242 0.0001 241	0.26851 0.0001 242	-0.28435 0.0001 242	0.32006 0.0001 242	0.06915 0.284 242	0.21702 0.0007 242	0.27898 0.0001 242	0.06854 0.2882 242	1.00000		
River	0.32431 0.0001 241	0.29848 0.0001 242	-0.22666 0.0004 242	0.26208 0.0001 242	0.03634 0.5737 242	0.30687 0.0001 242	-0.00695 0.9143 242	0.33739 0.0001 242	0.11768 0.0676 242	1.00000	
Relief	-0.05941 0.3585 241	-0.01225 0.8497 242	0.20966 0.001 242	-0.08425 0.1915 242	-0.21792 0.0006 242	0.3354 0.0001 242	0.09614 0.1359 242	-0.04714 0.4655 242	0.09474 0.1417 242	0.04824 0.455 242	1.00000

APPENDIX B: DESCRIPTIVE STATISTICS

Southwest Sector: Pearson Correlation Coefficients / PROB > |R| under H0: RHO=0 / N= Number of Observations

	D'70	D'80	Distance	Rail	H'way	Forest	Lake	N'Park	Golf	River	Relief
Density 1970	1.00000										
Density 1980	0.92509 0.0001 213	1.00000									
Distance	-0.68929 0.0001 213	-0.69663 0.0001 218	1.00000								
Rail	0.70363 0.0001 213	0.65478 0.0001 218	-0.69136 0.0001 218	1.00000							
Highway	0.39511 0.0001 213	0.34651 0.0001 218	-0.1702 0.0118 218	0.19821 0.0033 218	1.00000						
Forest	0.14422 0.0354 213	0.22851 0.0007 218	-0.33969 0.0001 218	0.24897 0.0002 218	-0.01533 0.822 218	1.00000					
Lake	-0.0752 0.2746 213	-0.01487 0.8272 218	0.11801 0.0821 218	-0.10447 0.1241 218	0.1485 0.0284 218	0.01353 0.8425 218	1.00000				
Nature Park	0.19995 0.0034 213	0.17069 0.0116 218	-0.1655 0.0144 218	0.17779 0.0085 218	-0.01338 0.8442 218	0.45569 0.0001 218	0.15965 0.0183 218	1.00000			
Golf	0.14192 0.0385 213	0.17285 0.0106 218	-0.2748 0.0001 218	0.17248 0.0107 218	-0.06204 0.3619 218	0.27713 0.0001 218	-0.00487 0.943 218	0.14239 0.0356 218	1.00000		
River	0.15392 0.0247 213	0.14496 0.0324 218	0.14183 0.0364 218	0.06378 0.3486 218	0.18879 0.0052 218	0.16302 0.016 218	0.16543 0.0145 218	0.10686 0.1157 218	-0.00604 0.9293 218	1.00000	
Relief	0.18662 0.0063 213	0.18919 0.0051 218	-0.22996 0.0006 218	0.15737 0.0201 218	-0.00394 0.9538 218	0.41296 0.0001 218	-0.18393 0.0065 218	0.10416 0.1252 218	0.19801 0.0033 218	0.22534 0.0008 218	1.00000

APPENDIX B: DESCRIPTIVE STATISTICS

South Sector: Pearson Correlation Coefficients / PROB > |R| under H0: RHO=0 / N= Number of Observations

	D70	D80	Distance	Rail	H'way	Forest	Lake	N'Park	Golf	River	Relief
Density 1970	1.00000										
Density 1980	0.96445 0.0001 97	1.00000									
Distance	-0.87332 0.0001 97	-0.89351 0.0001 97	1.00000								
Rail	0.66606 0.0001 97	0.64824 0.0001 97	-0.67907 0.0001 97	1.00000							
Highway	0.30917 0.0021 97	0.36826 0.0002 97	-0.48542 0.0001 97	0.18572 0.0686 97	1.00000						
Forest	0.12839 0.2101 97	0.21784 0.0321 97	-0.21667 0.033 97	0.0005 0.9961 97	0.10324 0.3143 97	1.00000					
Lake	0.30174 0.0027 97	0.32209 0.0013 97	-0.33912 0.0007 97	0.02532 0.8056 97	0.27551 0.0063 97	0.10316 0.3146 97	1.00000				
Nature Park	0.3213 0.0013 97	0.36226 0.0003 97	-0.30701 0.0022 97	0.16814 0.0997 97	0.13028 0.2034 97	0.44571 0.0001 97	0.21312 0.0361 97	1.00000			
Golf	0.29695 0.0031 97	0.2901 0.0039 97	-0.23312 0.0216 97	0.20464 0.0444 97	-0.01688 0.8696 97	0.20278 0.0464 97	0.14533 0.1555 97	0.03716 0.7178 97	1.00000		
River	0.4749 0.0001 97	0.44275 0.0001 97	-0.57242 0.0001 97	0.36725 0.0002 97	0.28831 0.0042 97	-0.0538 0.6007 97	0.30656 0.0023 97	0.14479 0.1571 97	0.06509 0.5264 97	1.00000	
Relief	-0.298 0.003 97	-0.27174 0.0071 97	0.34035 0.0006 97	-0.2229 0.0282 97	-0.34198 0.0006 97	0.34224 0.0006 97	-0.26306 0.0092 97	-0.06027 0.5576 97	0.19255 0.0588 97	-0.5298 0.0001 97	1.00000

APPENDIX B: DESCRIPTIVE STATISTICS

Suburban Region: Pearson Correlation Coefficients / PROB > |R| under HO: RHO=0 / N= Number of Observations

	D70	D80	Distance	Rail	H'way	Forest	Lake	N'Park	Golf	River	Relief
Density 1970	1.00000										
Density 1980	0.94472 0.0001 903	1.00000									
Distance	-0.70656 0.0001 903	-0.71333 0.0001 911	1.00000								
Rail	0.65626 0.0001 903	0.61769 0.0001 911	-0.49311 0.0001 913	1.00000							
Highway	0.2901 0.0001 903	0.27621 0.0001 911	-0.40638 0.0001 913	0.12316 0.0002 913	1.00000						
Forest	0.06266 0.0598 903	0.10477 0.0015 911	0.01826 0.5816 913	0.1242 0.0002 913	-0.13058 0.0001 913	1.00000					
Lake	0.09646 0.0037 903	0.11558 0.0005 911	0.08764 0.0081 913	0.0432 0.1922 913	-0.10081 0.0023 913	0.1685 0.0001 913	1.00000				
Nature Park	0.18556 0.0001 903	0.19135 0.0001 911	-0.23256 0.0001 913	0.09858 0.0029 913	0.12854 0.0001 913	0.26295 0.0001 913	0.03754 0.2572 913	1.00000			
Golf	0.2604 0.0001 903	0.27649 0.0001 911	-0.26288 0.0001 913	0.21915 0.0001 913	0.07611 0.0215 913	0.10556 0.0014 913	0.00272 0.9345 913	0.0448 0.1762 913	1.00000		
River	0.19229 0.0001 903	0.192 0.0001 911	-0.11964 0.0003 913	0.11088 0.0008 913	0.13995 0.0001 913	0.17 0.0001 913	-0.01676 0.613 913	0.1806 0.0001 913	0.0136 0.6815 913	1.00000	
Relief	-0.10637 0.0014 903	-0.06821 0.0396 911	0.28068 0.0001 913	-0.05284 0.1106 913	-0.24316 0.0001 913	0.30283 0.0001 913	0.01463 0.6589 913	-0.01787 0.5896 913	0.01496 0.6516 913	0.07732 0.0195 913	1.00000

BIBLIOGRAPHY

Abbott, C. "Necessary Adjuncts to Its Growth: The Railroad Suburbs of Chicago, 1854-1875." *Journal of the Illinois Historical Society* 73 (1980): 117-131.

Adams, J.S. "Directional Bias in Intra-Urban Migration." *Economic Geography* 45 (1969): 302-323.

_____. "Residential Structure of Midwestern Cities." *Annals of the Association of American Geographers* 60 (1970): 37-62.

Agnew, J.A.; Mercer, J.; and Sopher, D.E. Introduction to *The City in Cultural Context*, edited by J.A. Agnew, J. Mercer, and D.E. Sopher. Boston: Allen and Unwin, 1985.

Allen, I.L., ed. *New Towns and the Suburban Dream*. Port Washington, N.Y.: Kennikat, 1977.

Alonso, W. "The Population Factor and Urban Structure." In *The Prospective City*, edited by A.P. Solomon, ch. 1. Cambridge: Massachusetts Institute of Technology Press, 1980.

Alperovich, G. "Neighborhood Amenities and Their Impact on Density Gradients." *Annals of Regional Science* 14.2 (1980): 51-64.

Anderson, J.E. "The Changing Structure of a City: Temporal Changes in Cubic Spline Urban Density Patterns." *Journal of Regional Science* 25 (1985): 413-425.

Appleton, J. *The Experience of Landscape*. New York: Wiley, 1975.

_____. "Landscape Evaluation: The Theoretical Vacuum." *Transactions of the Institute of British Geographers* 66 (1975): 120-123.

Atkisson, A.A., and Robinson, I.M. "Amenity Resources for Urban Living." In *The Quality of the Urban Environment*, edited by H.S. Perloff, ch. 5. Baltimore: Johns Hopkins University Press Resources for the Future, 1969).

Entries preceded by * designate separate map sheets.

Balling, J.D., and Falk, J.H. "Development of Visual Preferences for Natural Environments." *Environment and Behavior* 14 (1982): 5-28.

Bartels, D. "Theorien nationaler Siedlungssysteme und Raumordnungspolitik." *Geographische Zeitschrift* 67 (1979): 110-146.

Bednarz, R.S. *The Effect of Air Pollution on Property Value in Chicago.* University of Chicago Department of Geography Research Paper no. 166. Chicago: University of Chicago Department of Geography, 1975.

Beesley, K.B., and Russwurm, L.H., eds. *The Rural-Urban Fringe: Canadian Perspectives*, York University Department of Geography, Geographical Monograph no. 10. Toronto: York University Department of Geography, 1981.

Berger, B.W. *Working-Class Suburb: A Study of Auto Workers in Suburbia.* Berkeley: University of California Press, 1971.

Berry, B.J.L. "Cities as Systems within Systems of Cities." In *Regional Development and Planning*, edited by J. Friedman and W. Alonso, ch. 6. Cambridge: Massachusetts Institute of Technology Press, 1964.

———. *The Human Consequences of Urbanization.* New York: St. Martin's, 1973.

———. "The Counterurbanization Process: How General?" In *Human Settlement Systems*, edited by N.M. Hansen, ch. 2. Cambridge, Mass.: Ballinger, 1978.

———, ed. *Urbanization and Counterurbanization.* Urban Affairs Annual Review no. 11. Newbury Park, Calif.: Sage Publications, 1976.

Berry, B.J.L., and Gillard, Q. *The Changing Shape of Metropolitan America: Commuting Patterns, Urban Fields, and Decentralization Processes.* Cambridge, Mass.: Ballinger, 1977.

Berry, B.J.L., and Horton, F.E. *Geographic Perspectives on Urban Systems.* Englewood Cliffs: Prentice Hall, 1970.

———. *Urban Environmental Management: Planning for Pollution Control.* Englewood Cliffs: Prentice Hall, 1974.

Berry, B.J.L., and Neils, E. "Location, Size, and Shape of Cities Influenced by Environmental Factors: The Urban Environment Writ Large." In *The Quality of the Urban Environment*, edited by H.S. Perloff, ch. 8. Washington, D.C.: Resources for the Future, 1969.

Berry, B.J.L. et al. "Urban Population Densities: Structure and Change." *Geographical Review* 53 (1963): 389-405.

Berry, B.J.L. et al. *Land Use, Urban Form, and Environmental Quality.* University of Chicago Department of Geography Research Paper no. 155. Chicago: University of Chicago Department of Geography, 1974.

Berry, B.J.L. et al. *Chicago: Transformation of an Urban System*. Association of American Geographers, Comparative Metropolitan Analysis Project. Cambridge, Mass.: Ballinger, 1976.

Berry, B.J.L. et al., *The Social Burdens of Environmental Pollution: A Comparative Metropolitan Data Source*. Cambridge, Mass.: Ballinger, 1977.

Binford, H.C. *The First Suburbs: Residential Communities on the Boston Periphery, 1815-1860*. Chicago: University of Chicago Press, 1985.

Borchert, J.R. "American Metropolitan Evolution." *Geographical Review* 57 (1967): 301-332.

_____. "America's Changing Metropolitan Regions." *Annals of the Association of American Geographers* 62 (1972): 352-373.

Bourne, L.S., ed. *Internal Structure of the City*. New York: Oxford University Press, 1971.

Briscoe, J.P. "Statistical Relationships of Growth in Urban Population Densities as Related to Selected Controllable Environmental Factors." Ph.D. diss., University of Louisville, 1977.

Brueckner, J.K. "A Note on Sufficient Conditions for Negative Exponential Population Densities." *Journal of Regional Science* 22 (1982): 353-359.

_____. "A Switching Regression Analysis of Urban Population Densities: Preliminary Results." *Papers of the Regional Science Association* 56 (1985): 71-87.

Bunce, M. "Rural Sentiment and the Ambiguity of the Urban Fringe." In *The Rural-Urban Fringe: Canadian Perspectives*, edited by K.B. Beesley and L.H. Russwurm. York University Department of Geography, Geographical Monograph no. 10. Toronto: York University Department of Geography, 1981, pp. 109-120.

Burgess, E.W. "The Growth of a City: An Introduction to a Research Project." In *The City*, 7th ed., edited by R.E. Park, pp. 47-62. Chicago: University of Chicago, 1974.

_____. "The Determination of Gradients in the Growth of the City." *Publications of the American Sociological Society* 21 (1927): 178-184.

Burgess, J., et al. "Exploring Environmental Values through the Medium of Small Groups, Part 2: Illustrations of a Group at Work." *Environment and Planning A* 20 (1988): 457-476.

Burgess, J.A. "Selling Places: Environmental Images for the Executive." *Regional Studies* 16 (1982): 1-17.

Cadwallader, M.T. "Neighborhood Evaluation in Residential Mobility." *Environment and Planning A* 11 (1979): 393-401.

———. "Migration and Intra-Urban Mobility." In *Population Geography: Progress and Prospect*, edited by M. Pacione, ch. 9. London: Croom Helm, 1986.

———. "Urban Geography and Social Theory." *Urban Geography* 9 (1988): 227-251.

Carruth, D.B. "Assessing Scenic Quality." *Landscape* 22 (1977): 31-34.

Carter, H. *An Introduction to Urban Historical Geography*. London: Arnold, 1983.

Cebula, R.J. *Determinants of Human Migration*. Lexington, Mass.: Lexington Books, 1979.

Chamberlin, E. *Chicago and Its Suburbs*. 1874. Reprint. New York: Arno Press, 1974.

Clark, C. "Urban Population Densities." *Journal of the Royal Statistical Society* 114 (1951): 490-496.

Clark, W.A.V. "Recent Research on Migration and Mobility." *Progress in Planning* 18 (1982): 1-56.

Clark, W.A.V., and Onaka, J.L. "Life Cycle and Housing Adjustment as Explanations of Residential Mobility." *Urban Studies* 20 (1983): 47-57.

Conzen, K.N. "Patterns of Residence in Early Milwaukee." In *The New Urban History*, edited by L.F. Schnore, ch. 5. Princeton: Princeton University Press, 1975.

Conzen, M.P. "Analytical Approaches to the Urban Landscape." In *Dimensions of Human Geography: Essays on Some Familiar and Neglected Themes*, edited by K.W. Butzer. University of Chicago Department of Geography Research Paper no. 186, pp. 134-158. Chicago: University of Chicago Department of Geography, 1978.

———. "The American Urban System in the Nineteenth Century." In *Geography and the Urban Environment: Progress in Research and Applications*, vol. 4, edited by D.T. Herbert and R.J. Johnston, pp. 316-320. London: Wiley and Sons, 1981.

———. "American Cities in Profound Transition: The New City Geography of the 1980s." *Journal of Geography* 82 (1983): 94-102.

Conzen, M.P., and Morales, M.J., eds. *Settling the Upper Illinois Valley: Patterns of Change in the I&M Canal Corridor, 1830-1900*. University of Chicago Committee on Geographical Studies, Studies on the I&M Canal Corridor no. 3. Chicago: University of Chicago Committee on Geographical Studies, 1989.

Conzen, M.R.G. *The Urban Landscape: Historical Development and Management: Papers by M.R.G. Conzen*, edited by J.W.R. Whitehand. Institute of British Geographers Special Publication no. 13. London: Academic Press, 1981.

Coppa, F.J. "Cities and Suburbs in Europe and the United States." In *Suburbia: The American Dream and Dilemma*, edited by P.C. Dolce, pp. 167-191. Garden City: Anchor Books, 1976.

Craik, K.H. "Appraising the Objectivity of Landscape Dimensions." In *Natural Environment*, edited by J.V. Krutilla, pp. 292-346. Baltimore: Johns Hopkins University Press Resources for the Future, 1972.

Cutter, S.L. *Rating Places: A Geographer's View on the Quality of Life*. Association of American Geographers Resource Publications in Geography. Washington, D.C.: Association of American Geographers, 1985.

de Borger, B. "Urban Population Density Functions: Some Belgian Evidence." *Annals of Regional Science* 8.3 (1979): 15-24.

de Crèvecœur, J.H. St.John. *Letters from an American Farmer*. London: Thomas Davies, 1782.

deVise, P. "The Geography of Wealth and Poverty in Suburban America: 1979 to 1985." City Club of Chicago, Chicago Regional Inventory Working Paper II.90 (March 1987). Photocopy.

Dear, M. "The Postmodern Challenge: Reconstructing Human Geography." *Transactions of the Institute of British Geographers* N.S. 13 (1988): 262-274.

Desbarats, J.M. "Spatial Choice and Constraints on Behavior." *Annals of the Association of American Geographers* 73 (1983): 340-357.

Diamond, D.B. "The Relationship between Amenities and Urban Land Prices." *Land Economics* 56 (1980): 21-32.

Diamond, D.B., and Tolley, G.S. "The Economic Roles of Urban Amenities." In *The Economics of Urban Amenities*, edited by D.B. Diamond and G.S. Tolley, ch. 1. New York: Academic Press, 1982.

Dillman, D.A. "Residential Preferences, Quality of Life, and the Population Turnaround." *American Journal of Agricultural Economics* 61 (1979): 960-966.

Dillman, D.A., and Dobash, R.P. *Preferences for Community Living and Their Implications for Population Distribution*. Washington State University Agricultural Experiment Station Research Bulletin no. 764. Spokane: Washington State University Agricultural Experiment Station, 1972.

Doney, R.S.; Evered, B.; and Kitchen, C.M. "Effects of Tree Conservation in the Urbanizing Fringe of Southern Ontario Cities, 1970-1984." *Urban Ecology* 9 (1986): 289-308.

Douglass, H. *The Suburban Trend.* 1925. Reprint. New York: Arno Press, 1970.

Drewnowski, J. *On Measuring and Planning the Quality of Life.* Publications of the Institute of Social Studies no. 11. The Hague: Mouton, 1974.

Dubin, R.A., and Sung, C.-H. "Spatial Variation in the Price of Housing: Rent Gradients in Non-monocentric Cities." *Urban Studies* 24 (1987): 193-204.

Duncan, B. "Variables in Urban Morphology." In *Urban Sociology,* edited by E.W. Burgess and D.J. Bogue, ch. 1. Chicago: University of Chicago Press, 1967.

Duncan, J.S., and Duncan, N.D. "A Cultural Analysis of Urban Residential Landscapes in North America: The Case of the Anglophile Elite." In *The City in Cultural Context,* edited by J.A. Agnew, J. Mercer, and D.E. Sopher, ch. 12. Boston: Allen and Unwin, 1985.

Dunn, E.S. Jr., *The Development of the U.S. Urban System,* vol. 1. Baltimore: Resources for the Future, 1983.

Ebner, M.H. "Re-reading Suburban America: Urban Population Deconcentration, 1810-1980." *American Quarterly* 37 (1985): 368-381.

Edmonston, B. *Population Distribution in American Cities.* Lexington, Mass.: Lexington Books, 1975.

Edmonston, B.; Goldberg, M.A.; and Mercer, J. "Urban Form in Canada and the United States: An Examination of Urban Density Gradients." *Urban Studies* 22 (1985): 209-217.

Elsom, D.M. "Pollution." In *Progress in Urban Geography,* edited by M. Pacione, ch. 10. London: Croom Helm, 1983.

Ermuth, F. *Residential Satisfaction and Urban Environmental Preferences.* York University Department of Geography, Geographical Monograph no. 3. Toronto: York University Department of Geography, 1974.

Fabos, J.G. et al. *Frederick Law Olmsted, Sr.* Amherst: University of Massachusetts Press, 1968.

Fielding, A.J. "Counterurbanization." In *Population Geography: Progress and Prospect,* edited by M. Pacione, ch. 8. London: Croom Helm, 1986.

Fuguitt, G.V., and Zuiches, J.J. "Residential Preferences and Population Distribution." *Demography* 12 (1975): 491-504.

Gans, H.J. *The Levittowners: Ways of Life and Politics in a New Suburban Community.* New York: Vintage Books, 1967.

General Soil Map of Illinois. Urbana-Champaign: University of Illinois Agricultural Experiment Station, 1982.

Gillard, Q. "The Effect of Environmental Amenities on Home Values: The Example of the View Lot." *Professional Geographer* 33 (1981): 216-220.

Glacken, C.J. "Reflections on the Man-Nature Theme as a Subject of Study." In *Future Environments of North America*, edited by F.F. Darling and J.P. Milton. Garden City: Natural History Press, 1966.

Gober, P., and Behr, M. "Central Cities and Suburbs as Distinct Places: Myth or Fact?" *Economic Geography* 58 (1982): 371-385.

Goheen, P.G. "Interpreting the American City." *Geographical Review* 64 (1974): 362-384.

Gold, J.R., and Burgess, J., eds. *Valued Environments*. London: Allen and Unwin, 1982.

Goldberg, M.A., and Mercer, J. *The Myth of the North American City*. Vancouver: University of British Columbia Press, 1986.

Gordon, D.M. "Capitalist Development and the History of American Cities." In *Marxism and the Metropolis*, edited by W.T. Tabb and L. Sawers, pp. 21-53. Oxford: Oxford University Press, 1980.

Gordon, P. "Deconcentration without a 'Clean Break.'" *Environment and Planning A* 11 (1979): 281-290.

Gottdiener, M. "Understanding Metropolitan Deconcentration: A Clash of Paradigms." *Social Science Quarterly* 64 (1983): 227-246.

Graves, P.E. "Migration and Climate." *Journal of Regional Science* 20 (1980): 227-237.

Graves, P.E., and Regulska, J. "Amenities and Migration over the Life-Cycle." In *The Economics of Urban Amenities*, edited by D.B. Diamond and G.S. Tolley, ch. 10. New York: Academic Press, 1982.

Greenhood, D. *Mapping*. 1944. Rev. ed. Chicago: University of Chicago Press, 1964.

Greenwood, M.J. "Human Migration: Theory, Models, and Empirical Studies." *Journal of Regional Science* 25 (1985): 521-544.

Griffith, D.A. "Evaluating the Transformation from a Monocentric to a Polycentric City." *Professional Geographer* 33 (1981): 189-196.

_____. "Modelling Urban Population Density in a Multi-centered City." *Journal of Urban Economics* 9 (1981a): 298-310.

Groop, R.E., and Muller, J.-C. "Evaluating an Urban Model: A Cartographic Approach." *American Cartographer* 5 (1978): 111-120.

Guest, A.M. "Population Suburbanization in American Metropolitan Areas, 1940-1970." *Geographical Analysis* 7 (1975): 267-283.

_____. "Patterns of Suburban Population Growth, 1970-75." *Demography* 16 (1979): 401-415.

Hadden, J.K. "Use of *ad hoc* Definitions." In *Sociological Methodology*, edited by E.F. Borgatta, pp. 276-285. San Francisco: Jossey-Bass, 1968.

Hadden, J.K., and Barton, J.J. "An Image That Will Not Die: Thoughts on the Anti-Urban Ideology." In *The Urbanization of the Suburbs*, edited by L.H. Masotti and J.K. Hadden, ch. 3. Newbury Park, Calif.: Sage Publications, 1973.

Haig, R.M. "Toward an Understanding of Metropolis, Parts I and II." *Quarterly Journal of Economics* 40 (1926): 179-208 and 402-434.

Hall, P. "Decentralization without End? A Re-evaluation." In *The Expanding City*, edited by J. Patten, pp. 125-155. London: Academic Press, 1983.

Harris, C.D. "Suburbs." *American Journal of Sociology* 49 (1943): 1-13.

Hart, J.F. "Population Change in the Upper Great Lakes." *Annals of the Association of American Geographers* 74 (1984): 221-243.

Hartke, W. "Die Sozialbrache als Phänomen der Geographischen Differenzierung der Landschaft." *Erdkunde* 10 (1956): 257-269.

Heaton, T. et al. "Residential Preferences, Community Satisfaction, and the Intention to Move." *Demography* 16 (1979): 565-573.

Hendrix, W.G., and Fabos, J.G. "Visual Land Compatibility as a Significant Contribution to Visual Landscape Quality." *International Journal of Environmental Studies* 8 (1975): 21-28.

Hepner, G.F. "An Analysis of Residential Developer Location Factors in a Fast Growing Urban Region." *Urban Geography* 4 (1983): 355-363.

Herzog, T.R. et al. "The Prediction of Preferences for Familiar Urban Places." *Environment and Behavior* 8 (1976): 627-645.

Hoch, I. "Variations in the Quality of Urban Life among Cities and Regions." In *Public Economics and the Quality of Life*, edited by L. Wingo and A. Evans, ch. 5. Baltimore: Johns Hopkins University Press, Resources for the Future and Centre for Environmental Studies, 1977.

Hoch, I., and Drake, J. "Wages, Climate, and the Quality of Life." *Journal of Environmental Economics and Management* 1 (1974): 284-295.

Hovinen, G.R. "Leapfrog Developments in Lancaster County: A Study of Residents' Perceptions and Attitudes." *Professional Geographer* 24 (1977): 194-199.

Hoyt, H. *One Hundred Years of Land Values in Chicago*. Chicago: University of Chicago Press, 1933.

———. *The Structure and Growth of Residential Neighborhoods in American Cities*. Washington, D.C.: U.S. Government Printing Office, 1939.

Hugill, P.J. "Home and Class among the American Landed Elite." In *The Power of Place,* edited by J.A. Agnew and J.S. Duncan, pp. 66-80. London: Unwin and Hyman, 1989.

Hurd, R.M. *Principles of City Land Values.* New York: Record and Guide, 1924.

Jackson, K.T. "Urban Deconcentration in the Nineteenth Century: A Statistical Inquiry." In *The New Urban History,* edited by L.F. Schnore, ch. 3. Princeton: Princeton University Press, 1975.

_____. *Crabgrass Frontier: The Suburbanization of the United States.* New York: Oxford University Press, 1985.

Jackson, P., and Smith, S.J. *Exploring Social Geography.* London: Allen and Unwin, 1984.

Johansson, P.-O. *The Economic Theory and Measurement of Environmental Benefits.* Cambridge: Cambridge University Press, 1987.

Johnson, H.B. *Order upon the Land: The U.S. Rectangular Land Survey and the Upper Mississippi Country.* New York: Oxford University Press, 1976.

Johnston, R.J. "Ideology and Quantitative Human Geography in the English-Speaking World." In *European Progress in Spatial Analysis,* edited by R.J. Bennett, ch. 2. London: Pion, 1981.

_____. *The American Urban System.* New York: St. Martin's, 1982.

_____. *City and Society: An Outline for Urban Geography.* London: Hutchinson, 1984.

Kahimbaara, J.A. "The Population Density Gradient and the Spatial Structure of a Third World City: Nairobi, a Case Study." *Urban Studies* 23 (1986): 307-322.

Kaplan, R. "The Green Experience." In *Humanscape: Environments for People,* edited by S. Kaplan and R. Kaplan, pp. 186-193. Ann Arbor: Ulrich's Books, 1982.

Kasarda, J.D., and Redfearn, G.V. "Differential Patterns of City and Suburban Growth in the United States." *Journal of Urban History* 2 (1975): 43-66.

Kim, J. "Factors Affecting Urban-to-Rural Migration." *Growth and Change* 14 (1983): 38-43.

Klein, D.C., ed. *Psychology of the Planned Community: The New Town Experience.* New York: Human Sciences Press, 1978.

Kohl, J.G. *Der Verkehr und die Ansiedelungen der Menschen in ihrer Abhängigkeit von der Gestaltung der Erdoberfläche.* Leipzig: Arnoldsche Buchhandlung, 1841.

Krakover, S. "Identification of Spatiotemporal Paths of Spread and Backwash." *Geographical Analysis* 15 (1983): 318-329.

Kurtz, R.A., and Eicher, J.B. "Fringe and Suburb: A Confusion of Concepts." *Social Forces* 37 (1958): 32-37.

Kutay, A. "Technological Change and Spatial Transformation in an Information Society, Part I: A Structural Model of Transition in an Urban System." *Environment and Planning A* 20 (1988): 569-593.

———. "Technological Change and Spatial Transformation in an Information Society, Part II: The Influence of New Information Technology on an Urban System." *Environment and Planning A* 20 (1988): 707-718.

Lamb, R.F. *Metropolitan Impacts on Rural America.* University of Chicago Department of Geography Research Paper no. 162. Chicago: University of Chicago Department of Geography, 1975.

Landforms of Illinois. Drawn by J.A. Bier. Urbana: Illinois State Geological Survey, Institute of National Resources, 1980.

Latham, R.F., and Yeates, M.H. "Population Density Growth in Metropolitan Toronto." *Geographical Analysis* 2 (1970): 177-185.

Leitner, H. "Urban Geography: Undercurrents of Change." *Progress in Human Geography* 11 (1987): 134-146.

Lewandowski, S.J. "The Built Environment and Cultural Symbolism in Post-Colonial Madras." In *The City in Cultural Context,* edited by J.A. Agnew, J. Mercer, and D.E. Sopher, ch. 11. Boston: Allen and Unwin, 1984.

Lewis, P.F. "Small Town in Pennsylvania." *Annals of the Association of American Geographers* 62 (1972): 323-351.

———. "The Galactic Metropolis." In *Beyond the Urban Fringe,* edited by R.H. Platt and G. Macinko, pp. 23-49. Minneapolis: University of Minnesota Press, 1983.

Lichter, D.T., and Fuguitt, G.V. "The Transition to Nonmetropolitan Population Deconcentration." *Demography* 19 (1982): 211-221.

Lin, G.Y. "Simple Markov Chain Model of Smog Probability in the South Coast Air Basin of California." *Professional Geographer* 33 (1981): 228-236.

Linneman, P. "Hedonic Prices and Residential Location." In *Economics of Urban Amenities,* edited by D.B. Diamond and G.S. Tolley, ch. 3. New York: Academic Press, 1982.

Linton, D.L. "The Assessment of Scenery as a Natural Resource." *Scottish Geographical Magazine* 84 (1969): 219-238.

Liu, B.C. *Quality of Life Indicators in U.S. Metropolitan Areas: A Statistical Analysis.* New York: Praeger, 1976.

London, B., and Flanagan, W.G. "Comparative Cultural Ecology: A Summary of the Field." In *The City in Comparative Perspective,* edited by J. Walton and L.H. Masotti, pp. 41-66. New York: Wiley, 1976.

MacDonald, J.F., and Bowman, H.W. "Some Tests of Alternative Urban Population Density Functions." *Journal of Urban Economics* 3 (1976): 242-252.

McHarg, I. *Design with Nature.* Garden City, N.Y.: Natural History Press, 1969.

McKenzie, R.D. "The Rise of Metropolitan Communities." 1933. Reprint. In *On Human Ecology: Selected Writings,* edited by A.H. Hawley, ch. 14. Chicago: University of Chicago Press, 1968.

Marchand, B. "Urban Growth Models Revisited: Cities as Self-Organizing Systems." *Environment and Planning A* 16 (1984): 949-964.

_____. *The Emergence of Los Angeles.* London: Pion, 1986.

Marsh, M.S., and Kaplan, S. "The Lure of the Suburbs." In *Suburbia: The American Dilemma,* edited by P.C. Dolce, pp. 37-58. Garden City: Anchor Books, 1976.

Marx, L. *The Machine in the Garden: Technology and the Pastoral Ideal in America.* New York: Oxford University Press, 1964.

_____. "The Puzzle of Antiurbanism in Classic American Literature." In *Cities of the Mind,* edited by L. Rodwin and R.M. Hollister, ch. 10. New York: Plenum Press, 1984.

Mayer, H.M. "Chicago als Weltstadt." In *Zum Problem der Weltstadt,* edited by J.H. Schultze, pp. 83-111. Berlin: De Gruyter, 1959.

_____. "A Survey of Urban Geography." In *The Study of Urbanization,* edited by P.M. Hauser and L.F. Schnore, ch. 3. New York: Wiley, 1965.

Mayer, H.M., and Wade, R.C. *Chicago: Growth of a Metropolis.* Chicago: University of Chicago Press, 1969.

Michelson, W. *Environmental Choice, Human Behavior, and Residential Satisfaction.* New York: Oxford University Press, 1977.

Mills, E.S. "Urban Density Functions." *Urban Studies* 7 (1970): 5-20.

_____. *Studies in the Structure of the Urban Economy.* Resources for the Future. Baltimore: Johns Hopkins University, 1972.

Mitchell, J.K. "Adjustment to New Physical Environments beyond the Metropolitan Fringe." *Geographical Review* 66 (1976): 18-31.

Mohl, R.A. "New Perspectives on American Urban History." In *The Making of Urban America*, edited by R.A. Mohl, pp. 293-316. Wilmington, Del.: Scholarly Resources, 1988.

Moriarty, B.M. "Socioeconomic Status and Residential Locational Choice." *Environment and Behavior* 6 (1974): 448-469.

Morrill, R.L. "Bases for Peripheral Urban Growth." In *Impact of Urbanization and Industrialization on the Landscape*, edited by D.R. Deskins. University of Michigan Department of Geography, Michigan Geographical Publication no. 25, pp. 95-117. Ann Arbor: University of Michigan Department of Geography, 1980.

Morrison, P.A., and Wheeler, J.P. "The Image of 'Elsewhere' in the American Tradition of Migration." In *Human Migration*, edited by W.H. McNeill and R.S. Adams, pp. 95-117. Bloomington: Indiana University Press, 1978.

Mudrak, L.Y. "Sensory Mapping and Preferences for Urban Nature." *Landscape Research* 7 (1982): 2-8.

──────. "Urban Residents' Landscape Preferences: A Method for Their Assessment." *Urban Ecology* 7 (1983): 91-123.

Mueller, C.F. *The Economics of Labor Migration: A Behavioral Analysis*. New York: Academic Press, 1982.

Müller-Wille, W. *Westfalen: Landschaftliche Ordnung und Bindung eines Landes*. Münster: Aschendorffsche Verlagsbuchhandlung, 1952.

Muller, P.O. *The Outer City: The Geographical Consequences of the Urbanization of the Suburbs*. Association of American Geographers Resource Paper no. 75-2. Washington, D.C.: Association of American Geographers, 1976.

──────. *Contemporary Suburban America*. Englewood Cliffs: Prentice Hall, 1981.

Muth, R.F. *Cities and Housing*. Chicago: University of Chicago Press, 1969.

Nelson, K.P. "Urban Economic and Demographic Change." *Research in Urban Economics* 4 (1984): 25-49.

Newling, B. "The Spatial Variation of Urban Population Densities." *Geographical Review* 59 (1969): 242-253.

Nicholson-Lord, D. *The Greening of the Cities*. London: Routledge and Kegan Paul, 1987.

* *Northeastern Illinois Basemap*, 1:126,720. Rev. ed. Chicago: Rand McNally and Co., 1982.

Northeastern Illinois Planning Commission. *Suburban Factbook*. Chicago: Northeastern Illinois Planning Commission, 1971.

_____. *Regional Data Report.* Chicago: Northeastern Illinois Planning Commission, 1978.

Numrich, R.P. *Essays on the Econometric Estimation of Urban Spatial Structure.* Ph.D. diss., State University of New York at Albany, 1986.

O'Connor, C.A. *A Sort of Utopia: Scarsdale, 1881-1981.* Albany: State University of New York Press, 1983.

Okabe, A., and Masuda, S. "Qualitative Analysis of Two-Dimensional Urban Population Distributions in Japan." *Geographical Analysis* 16 (1984): 301-312.

Orishimo, I. *Urbanization and Environmental Quality.* Boston: Kluwer Nijhoff, 1982.

Pacione, M. "Revealed Preferences and Residential Environmental Quality." In *Quality of Life and Human Welfare,* edited by M. Pacione and G. Gordon, ch. 5. Norwich, U.K.: Geog Books, 1984.

Pahl, R.E. *Urbs in Rure: The Metropolitan Fringe in Herfordshire.* London School of Economics and Political Sciences Geographical Paper no. 2. London: London School of Economics and Political Sciences, 1964.

Palm, R. *The Geography of American Cities.* New York: Oxford University Press, 1981.

Parr, J.B. "A Population-Density Approach to Regional Spatial Structure." *Urban Studies* 22 (1985): 289-303.

Patel, D.I. *Exurbs: Urban Residential Development in the Countryside.* Washington, D.C.: University Press of America, 1980.

Perin, C. *Everything in Its Place: Social Order and Land Use in America.* Princeton: Princeton University Press, 1977.

Perkins, H.C. "Bulldozers in the Southern Part of Heaven, Parts I and II." *Environment and Planning A* 20 (1988): 285-308 and 435-456.

Perloff, H.S. "A Framework for Dealing with the Urban Environment: Introductory Statement." In *The Quality of the Urban Environment,* edited by H.S. Perloff, ch. 1. Resources for the Future. Baltimore: Johns Hopkins Press, 1969.

Perloff, H.S., and Wingo, L. Jr. "Natural Resource Endowment and Regional Economic Growth." In *Regional Development and Planning,* edited by J. Friedman and W. Alonso, ch. 11. Cambridge: Massachusetts Institute of Technology Press, 1964.

Perloff, H.S. et al. *Regions, Resources, and Economic Growth.* Resources for the Future. Baltimore: Johns Hopkins Press, 1960.

Pfeiffer, G. "...und man sollte J.G. Kohl nicht vergessen!" In *Mensch und Erde.* Institut für Geographie und Länderkunde Geographische Kommission für Westfalen Westfälische Geographische Studien no.

33, pp. 221-236. Münster: Institut für Geographie und Länderkunde Geographische Kommission für Westfalen, 1976.

Platt, R.H. *The Open Space Decision Process.* University of Chicago Department of Geography Research Paper no. 142. Chicago: University of Chicago, Department of Geography, 1972.

Pole, J.R. *American Individualism and the Promise of Progress.* Oxford: Clarendon Press, 1980.

Pollard, R. "View Amenities, Building Heights, and Housing Supply." In *The Economics of Urban Amenities,* edited by D.B. Diamond and G.S. Tolley, ch. 5. New York: Academic Press, 1982.

Porteous, J.D. "Urban Environmental Aesthetics." In *Environmental Aesthetics: Essays in Interpretation,* edited by B. Sadler and A. Carlson, ch. 4. University of Victoria Department of Geography, Western Geographical Series no. 20. Victoria: University of Victoria Department of Geography, 1982.

Pred, A. *Urban Growth and City-Systems in the United States, 1840-1860.* Cambridge, Mass.: Harvard University Press, 1980.

Preston, V. "A Multidimensional Scaling Analysis of Individual Differences in Residential Area Evaluation." *Geografiska Annaler B* 64 (1982): 17-26.

Pryor, R.J. "Defining the Rural Urban Fringe." In *Internal Structure of the City,* edited by L.S. Bourne, pp. 59-68. New York: Oxford University Press, 1972.

Quaternary Deposits of Illinois. Compiled by J.A. Lineback. Springfield: Illinois State Geological Survey, 1979.

Quigley, J.M., and Weinberg, D.H. "Intra-Urban Residential Mobility: A Review and Synthesis." *International Regional Science Review* 2 (1977): 41-66.

Rasmussen, P., and Giagnorio, C., eds. *Environmentally Sensitive Features File Code Book and Definitions of Variables.* Northeastern Illinois Planning Commission Project no. 0501.01. Chicago: Northeastern Illinois Planning Commission, 1979.

Ratzel, F. *Erdenmacht und Völkerschicksal.* Edited by K. Haushofer. Stuttgart: A. Kröner, 1940.

Reps, J.W. *Cities in the American West: A History of Frontier Urban Planning.* Princeton: Princeton University Press, 1979.

———. *The Forgotten Frontier: Urban Planning in the American West.* Columbia: University of Missouri Press, 1981.

Richardson, H.W. "On the Possibility of Positive Rent Gradients." *Journal of Urban Economics* 4 (1977): 246-257.

_____. *Urban Economics*. Hinsdale: Dryden Press, 1978.

Richardson, J.F. "The Dynamics of American Urban Development." In *Cities in the Twenty-first Century*, edited by G. Gappert and R.V. Knight, ch. 2. Newbury Park, Calif.: Sage Publications, 1982.

Rodriguez-Bachiller, A. "Discontiguous Urban Growth and the New Urban Economics: A Review." *Urban Studies* 23 (1986): 79-104.

Roseman, C.C. "Exurban Areas and Exurban Migration." In *The American Metropolitan System*, edited by S.D. Brunn and J.O. Wheeler, ch. 4. New York: Wiley, 1980.

_____. "Cartographic Analysis of Population Change." *Annals of the Association of American Geographers* 75 (1985): 133-135.

Ruppert, K. "Zur Definition des Begriffes 'Sozialbrache.'" *Erdkunde* 11 (1957): 226-231.

Sack, R.D. *Human Territoriality: Its Theory and History*. Cambridge: Cambridge University Press, 1986.

Sadler, B., and Carlson, A. "Environmental Aesthetics in Interdisciplinary Perspective." In *Environmental Aesthetics*, edited by B. Sadler and A. Carlson. University of Victoria Department of Geography, Western Geographical Series vol. 20, ch. 1. Victoria, B.C.: University of Victoria Department of Geography, 1982.

Schaffer, D. *Garden Cities for America: The Radburn Experience*. Philadelphia: Temple University Press, 1982.

Schmid, J.A. "The Environmental Impact of Urbanization." In *Perspectives on Environment*, edited by M.W. Mikesell and I.R. Manners, ch. 8. Commission on College Geography. Washington, D.C.: Association of American Geographers, 1974.

_____. *Urban Vegetation*. University of Chicago Department of Geography Research Paper no. 161. Chicago: University of Chicago Department of Geography, 1975.

Schnore, L.F. "The Timing of Metropolitan Decentralization." *Journal of the American Institute of Planners* 25 (1959): 200-206.

Shafer, E.L., and Hamilton, J.E. "Natural Landscape Preferences: A Predictive Model." *Journal of Leisure Research* 1 (1969): 1-19.

Shlay, A.B. "Taking Apart the American Dream: The Influence of Income and Family Composition on Residential Evaluations." *Urban Studies* 23 (1986): 253-270.

Simmons, J.W. "Changing Residence in the City: A Review of Intraurban Mobility." *Geographical Review* 58 (1968): 622-651.

_____. "The Organization of the Urban System." In *Systems of Cities*, edited by L.S. Bourne and J.W. Simmons, pp. 61-69. New York: Oxford University Press, 1978.

Smith, V.K. "Residential Location and Environmental Amenities." *Regional Studies* 11 (1977): 47-61.

Sofranko, A.J. "Urban Migrants to the Rural Midwest: Some Understandings and Misunderstandings." In *Population Redistribution in the Midwest*, edited by C.C. Roseman et al., ch. 5. North Central Regional Center for Rural Development. Ames: Iowa State University Press, 1981.

Sofranko, A.J., and Fliegel, F.C. "Neglected Components of Rural Population Growth." *Growth and Change* 14 (1983): 42-49.

Sonnenfeld, J. "Variable Values in Space and Landscape." *Journal of Social Issues* 22.4 (1966): 71-82.

Stanislawski, D. "The Origin and Spread of the Grid-Pattern Town." *Geographical Review* 36 (1946): 105-120.

Stilgoe, J.R. *Borderland: Origins of the American Suburb, 1820-1939*. New Haven: Yale University Press, 1988.

Stoneall, L. *Country Life, City Life*. New York: Praeger, 1983.

Tarr, J.A. "From City to Suburb." In *American Urban History*, edited by A.B. Callow Jr., pp. 202-212. New York: Oxford University Press, 1973.

Taylor, P.J. *Quantitative Methods in Geography*. London: Houghton and Mifflin, 1977.

Tomioka, S., and Tomioka, E.M. *Planned Unit Developments: Design and Regional Impact*. New York: Wiley, 1984.

Tuan, Y.F. *Topophilia*. Englewood Cliffs: Prentice Hall, 1974.

Ullman, E.L. "Amenities as a Factor in Regional Growth." *Geographical Review* 44 (1954): 119-132.

Ulrich, R.S. "Visual Landscape Preferences: A Model and Application." *Man-Environment Systems* 7 (1977): 279-293.

U.S. Bureau of the Census. *1980 Census of Population, PC80-1-B15*. Washington, D.C.: Government Printing Office, 1982.

_____. *1980 General Population Characteristics: United States Summary*. Washington, D.C.: U.S. Government Printing Office, 1982.

_____. *1980 User's Guide, Part B*. Washington, D.C.: Government Printing Office, 1982.

_____. *Census of Governments*. Washington, D.C.: Government Printing Office, 1982.

Vance, J.E. "California and the Search for the Ideal." *Annals of the Association of American Geographers* 62 (1972): 185-210.

———. *This Scene of Man: The Role and Structure of the City in the Geography of Western Civilization.* New York: Harper and Row, 1977.

Vining, D.R., and Kontuly, T. "Population Dispersal from Major Metropolitan Regions: An International Comparison." *International Regional Science Review* 3 (1978): 49-73.

Vining, D.R.; Kontuly, T.; and Strauss, A. "A Demonstration That the Current Deconcentration of Population in the U.S. Is a Clean Break with the Past." *Environment and Planning A* 9 (1977): 751-758.

Walker, R.A. "The Transformation of Urban Structure in the Nineteenth Century and the Beginnings of Suburbanization." In *Urbanization and Conflict in Market Societies*, edited by K.R. Cox, ch. 8. Chicago: Marouffa, 1978.

Ward, D. "The Place of Victorian Cities in Development Approaches to Urbanization." In *The Expanding City*, edited by J. Patten, ch. 12. London: Academic Press, 1983.

Wardwell, J.M. "Toward a Theory of Rural-Urban Migration in the Developed World." In *New Directions in Urban-Rural Migration,* edited by D.L. Brown and J.M. Wardwell, ch. 4. London: Academic Press, 1980.

Watson, J.W. "The Image of Nature in America." In *The American Environment*, edited by J.W. Watson and T. O'Riordan, ch. 5. New York: Wiley, 1976.

———. *Social Geography of the United States*. London: Longman, 1979.

Webber, M.M. "The Urban Place and the Nonplace Urban Realm." In *Explorations into Urban Structure*, edited by M.M. Webber et al., ch. 2. Philadelphia: University of Pennsylvania Press, 1964.

Weichhart, P. "Assessment of the Natural Environment: A Determinant of Residential Preferences." *Urban Ecology* 7 (1982/83): 325-342.

Weightman, B.A. "Arcadia in Suburbia: Orange County, California." *Journal of Cultural Geography* 2 (1981): 55-69.

Weinstein, N.D. "The Statistical Prediction of Environmental Preferences: Problems of Validity and Application." *Environment and Behavior* 4 (1976): 611-626.

White, S.E. "The Influence of Urban Residential Preference on Spatial Behavior." *Geographical Review* 71 (1981): 176-187.

Whitehand, J.W.R. "Taking Stock of Urban Geography." *Area* 18 (1986): 147-151.

———. "Urban Morphology." In *Historical Geography: Progress and Prospect*, edited by M. Pacione, ch. 9. London: Croom Helm, 1987.

Whyte, W.H. *The Organization Man*. Garden City, N.Y.: Anchor Books, 1956.

Williams, J.D., and McMillan, D.B. "Migration Decision Making among Nonmetropolitan-Bound Migrants." In *New Directions in Rural-Urban Migration*, edited by D.L. Brown and J.M. Wardwell, ch. 8. New York: Academic Press, 1980.

Wingo, L. "Objective, Subjective, and Collective Dimensions of the Quality of Life." In *Public Economics and the Quality of Life*, edited by L. Wingo and A. Evans, ch. 1. Resources for the Future and Centre for Environmental Studies. Baltimore: Johns Hopkins University Press, 1977.

Winsborough, H.H. "City Growth and City Structure." *Journal of Regional Science* 4 (1962): 35-49.

———. "An Ecological Approach to the Theory of Suburbanization." *American Journal of Sociology* 63 (1963): 565-570.

Wood, R.C. "The American Suburb." In *Man and the Modern City*, edited by E. Green et al., pp. 112-121. Pittsburgh: University of Pittsburgh Press, 1963.

Woods, R.I. "Theory and Methodology in Population Geography." In *Population Geography: Progress and Prospect*, edited by M. Pacione, ch. 2. London: Croom Helm, 1986.

Zehner, R.B. *Indicators of the Quality of Life in New Communities*. Cambridge, Mass.: Ballinger, 1977.

Zeigler, D.J., and Brunn, S.D. "Geopolitical Fragmentation and the Pattern of Growth and Need: Defining the Cleavage between Sunbelt and Frostbelt Metropolises." In *The American Metropolitan System*, edited by S.D. Brunn and J.O. Wheeler, ch. 6. New York: Wiley, 1980.

Zelinsky, W. *The Cultural Geography of the United States*. Foundations in Geography Series. Englewood Cliffs: Prentice Hall, 1973.

———. "Coping with the Migration Turnaround: The Theoretical Challenge." *International Regional Science Review* 2 (1977): 175-177.

Zikmund, J. II, and Dennis, D.E. *Suburbia: A Guide to Information Sources*. Urban Studies Information Guide Series, vol. 9. Detroit: Gales Research Co., 1979.

Zube, E.H. "Rating Everyday Rural Landscapes of the Northeastern United States." *Landscape Architecture* 63 (1973): 370-375.

———. *Environmental Evaluation: Perception and Public Policy*. Monterey, Calif.: Brooks-Cole, 1980.

Zube, E.H. et al. *Landscape Assessment: Values, Perceptions, and Resources*. Stroudsburg, Pa.: Dowden, Hutchinson, and Ross, 1975.

INDEX

accessibility 41
Adams, J.S. 35
aerial photographs 45
amenities 26, 41 *See also* Externalities.
    as pull factors 32
    life span 120
amenity-rich areas 116, 120
antiurban bias 27
Arcadia 30
Arlington 74
attitude-discrepant behavior 20
Aurora 81, 93, 106
auto-air-amenity epoch 34

Bartels, D. 34
Beecher 89
beltline 105
Berry, B.J.L. 32
Borchert, J.R. 34

Calumet Sag Channel 85
Chicago Heights 93
Chicago metropolitan region
    sectors 61
Chicago-Milwaukee-St.Paul and Pacific Railroad 68
Chicago-Northwestern Railroad 64, 68, 74
Chicago Outer Belt Line 81, 87
Chicago Sanitary & Ship Canal 85
city in the garden 9, 29
city-suburb dichotomy 12
Clark, C. 42
clustered neighborhood design 78
compactness of suburban settlement 90
competitive advantage 15
Cook County 74, 79
counter-culture 21

counterurbanization 32
country retreats 68
Crystal Lake 75
cubic spline functions 40
cultural competition 21
cultural context 21-22
cultural factors 27
cultural predispositions 27
cultural traits 22

de Crèvecœur, J.H. St.John 32
density gradient 38
    central density 38
    central density crater 43
    crest 43
Des Plaines River 66, 68, 81, 82, 85
disamenities. *See* Externalities.
distance bands 112
DuPage County 79
Du Page River 85

Eastwest Tollway 79
Edens Expressway 64
Eisenhower Expressway 79
Elgin 81, 93, 106
Elmhurst 79
emergent features 119
Evanston 64
externalities 17, 26, 41
exurban communities 105, 111
exurban periphery 116
exurbia 29
exurbs 68

factory towns 106
featureless plain 39
field surveys 45

floodplains 66
forest preserves 66
forests 66, 68, 88
Fox River 81, 82
Fox River valley 75, 77
frictionless city 16

garden city 28
geographic information system 121
geomorphic city 16
Gillard, Q. 32
golf courses 78
gross population densities 60

Harvard 74
Hebron 64
hegemonic culture 21
highway access 88
highway ribbon 64
Hoyt, H. 25
human ecology 19
Hurd, R.M. 25

Illinois Central railroad 89
Indian treaty boundary 47
individualism 27
infilling 105
inundated bottomlands 66
intrametropolitan accessibility
　　measures 57
isoplethic maps 102

Joliet 84, 93, 106

Kane County 79

Lake Barrington 77
Lake County 64, 71
lake district 64
Lake Forest 64
Lake in the Hills 77
Lake Michigan 69
lakes 65
landscape
　　beneficent 30
　　cultural 11
　　glacially overformed 2
　　physiographic 2
　　physiographically bland 93
　　secular 29
　　suburban 9

urban 9, 22
valued 18
landscape components
　　emergent 95
　　fixtures 94
　　transitional 94
　　typology 94
landscape fixtures 118
landscape preferences 30
land-use theory 40
Lemont 87
Libertyville 64
locational competition 21
locational preference 20
Lockport 87

McHenry County 64, 74
mechanistic view of the world 29
megalopolis 30
metaphors of American culture 29
metropolitan evolution 34
metropolitan growth
　　reinterpretation 120
models of urban systems
　　development and organization 33
multiple-regression model 41-42
　　forward-selection technique 42

national city systems 15
Natural Landscape Components 49
natural landscape feature
　　inferred attraction 9
　　suburban development 10
nature parks 88
neoclassical economics 19
net population densities 60
Northeastern Illinois Planning
　　Commission 3, 44
north-shore margin 69
Northwest Tollway 75

Oakwood Hills 77
O'Hare International Airport 74
Orland Park 84, 88

Palos Hills 88
Park Ridge 74
pattern of residential growth 35
planned unit development 78
pollution 17
polycentric metropolitan areas 40

population change
    absolute 106
    relative 106
population density
    alternative density functions 40
    concentric rings 35
    negative exponential function 37
proxy measures 33
public goods 26

quality of life 18

radial density ridges 105
rectangular land survey 45
relative relief 65
residential relocation 20
Riverside 81
romantic suburb 82
rural Eden 29, 43

satellite cities 81
satellite images 45
scales of study 23
scenic quality 30
Schaumburg 79
sectoral division 118
sectorality 62
sectoral limits 63
Simmons, J.W. 33
Skokie River 66
social areas 22
social theory 21
soil association 2
spatial aggregation 47
spatial association 117
spatial matrix 45
spatial mobility 28
spatial series of equilibrium conditions 39
spread-backwash processes 111
suburb
    definition of 11
    physical categories 11
    political fragmentation 12, 28
    residential ideal 7
    settlement sequence 68
suburban ambience 10
suburban dream 97
suburban expansion 78
    inner ring 105
    outer ring 105
    wave-like progression 111

suburbanization
    compactness 38
    decentralization 14
    deconcentration 14, 38
    diagnostic variables 14
suburbia
    as haven 8
    as promise 8
    as synthesis 8
systems of cities 14

themes of American culture 32
topographic variation 82
township-range system 29
transitional features 118
Tristate Tollway 64, 68, 75, 79, 84, 89

U.S. Public Land Survey 44
Ullman, E.L. 25
urban form
    environmental factors 18
    one-dimensional 39
    presumptive conditions 39
urban growth
    historic sequence 35
    transport eras 35
urban models
    behavioral-perceptual 19
    institutional-managerial 19
    Marxist 19
    population density gradients 23
    positivist 20
    process-oriented theories 19
    theoretical incompatibility 23
urban rent gradients 39
urban structure
    major themes 19
    rational economic behavior 36
    site quality 102

Vance, J.E. 29
vegetational division 2

Waukegan 93, 106
wave analog model 111
Winthrop Harbor 64

Zelinsky, W. 27

THE UNIVERSITY OF CHICAGO
GEOGRAPHY RESEARCH PAPERS
(Lithographed, 6 x 9 inches)

Titles in Print

127. GOHEEN, PETER G. *Victorian Toronto, 1850 to 1900: Pattern and Process of Growth.* 1970. xiii + 278 p.
130. GLADFELTER, BRUCE G. *Meseta and Campina Landforms in Central Spain: A Geomorphology of the Alto Henares Basin.* 1971. xii + 204 p.
131. NEILS, ELAINE M. *Reservation to City: Indian Migration and Federal Relocation.* 1971. x + 198 p.
132. MOLINE, NORMAN T. *Mobility and the Small Town, 1900-1930.* 1971. ix + 169 p.
133. SCHWIND, PAUL J. *Migration and Regional Development in the United States, 1950-1960.* 1971. x + 170 p.
134. PYLE, GERALD F. *Heart Disease, Cancer and Stroke in Chicago: A Geographical Analysis with Facilities, Plans for 1980.* 1971. ix + 292 p.
136. BUTZER, KARL W. *Recent History of an Ethiopian Delta: The Omo River and the Level of Lake Rudolf.* 1971. xvi + 184 p.
139. McMANIS, DOUGLAS R. *European Impressions of the New England Coast, 1497-1620.* 1972. viii + 147 p.
140. COHEN, YEHOSHUA S. *Diffusion of an Innovation in an Urban System: The Spread of Planned Regional Shopping Centers in the United States, 1949-1968.* 1972. ix + 136 p.
141. MITCHELL, NORA. *The Indian Hill-Station: Kodaikanal.* 1972. xii + 199 p.
142. PLATT, RUTHERFORD H. *The Open Space Decision Process: Spatial Allocation of Costs and Benefits.* 1972. xi + 189 p.
143. GOLANT, STEPHEN M. *The Residential Location and Spatial Behavior of the Elderly: A Canadian Example.* 1972. xv + 226 p.
144. PANNELL, CLIFTON W. *T'ai-Chung, T'ai-wan: Structure and Function.* 1973. xii + 200 p.
145. LANKFORD, PHILIP M. *Regional Incomes in the United States, 1929-1967: Level, Distribution, Stability, and Growth.* 1972. x + 137 p.
146. FREEMAN, DONALD B. *International Trade, Migration, and Capital Flows: A Quantitative Analysis of Spatial Economic Interaction.* 1973. xiv + 201 p.
147. MYERS, SARAH K. *Language Shift among Migrants to Lima, Peru.* 1973. xiii + 203 p.
148. JOHNSON, DOUGLAS L. *Jabal al-Akhdar, Cyrenaica: An Historical Geography of Settlement and Livelihood.* 1973. xii + 240 p.
149. YEUNG, YUE-MAN. *National Development Policy and Urban Transformation in Singapore: A Study of Public Housing and the Marketing System.* 1973. x + 204 p.
150. HALL, FRED L. *Location Criteria for High Schools: Student Transportation and Racial Integration.* 1973. xii + 156 p.
151. ROSENBERG, TERRY J. *Residence, Employment, and Mobility of Puerto Ricans in New York City.* 1974. xi + 230 p.
152. MIKESELL, MARVIN W., ed. *Geographers Abroad: Essays on the Problems and Prospects of Research in Foreign Areas.* 1973. ix + 296 p.
153. OSBORN, JAMES. *Area, Development Policy, and the Middle City in Malaysia.* 1974. x + 291 p.

154. WACHT, WALTER F. *The Domestic Air Transportation Network of the United States.* 1974. ix + 98 p.
155. BERRY, BRIAN J. L. et al. *Land Use, Urban Form and Environmental Quality.* 1974. xxiii + 440 p.
156. MITCHELL, JAMES K. *Community Response to Coastal Erosion: Individual and Collective Adjustments to Hazard on the Atlantic Shore.* 1974. xii + 209 p.
157. COOK, GILLIAN P. *Spatial Dynamics of Business Growth in the Witwatersrand.* 1975. x + 144 p.
160. MEYER, JUDITH W. *Diffusion of an American Montessori Education.* 1975. xi + 97 p.
162. LAMB, RICHARD F. *Metropolitan Impacts on Rural America.* 1975. xii + 196 p.
163. FEDOR, THOMAS STANLEY. *Patterns of Urban Growth in the Russian Empire during the Nineteenth Century.* 1975. xxv + 245 p.
164. HARRIS, CHAUNCY D. *Guide to Geographical Bibliographies and Reference Works in Russian or on the Soviet Union.* 1975. xviii + 478 p.
165. JONES, DONALD W. *Migration and Urban Unemployment in Dualistic Economic Development.* 1975. x + 174 p.
166. BEDNARZ, ROBERT S. *The Effect of Air Pollution on Property Value in Chicago.* 1975. viii + 111 p.
167. HANNEMANN, MANFRED. *The Diffusion of the Reformation in Southwestern Germany, 1518-1534.* 1975. ix + 235 p.
168. SUBLETT, MICHAEL D. *Farmers on the Road: Interfarm Migration and the Farming of Noncontiguous Lands in Three Midwestern Townships. 1939-1969.* 1975. xiii + 214 p.
169. STETZER, DONALD FOSTER. *Special Districts in Cook County: Toward a Geography of Local Government.* 1975. xi + 177 p.
171. SPODEK, HOWARD. *Urban-Rural Integration in Regional Development: A Case Study of Saurashtra, India—1800-1960.* 1976. xi + 144 p.
172. COHEN, YEHOSHUA S., and BRIAN J. L. BERRY. *Spatial Components of Manufacturing Change.* 1975. vi + 262 p.
173. HAYES, CHARLES R. *The Dispersed City: The Case of Piedmont, North Carolina.* 1976. ix + 157 p.
174. CARGO, DOUGLAS B. *Solid Wastes: Factors Influencing Generation Rates.* 1977. 100 p.
176. MORGAN, DAVID J. *Patterns of Population Distribution: A Residential Preference Model and Its Dynamic.* 1978. xiii + 200 p.
177. STOKES, HOUSTON H.; DONALD W. JONES; and HUGH M. NEUBURGER. *Unemployment and Adjustment in the Labor Market: A Comparison between the Regional and National Responses.* 1975. ix + 125 p.
180. CARR, CLAUDIA J. *Pastoralism in Crisis. The Dasanetch and Their Ethiopian Lands.* 1977. xx + 319 p.
181. GOODWIN, GARY C. *Cherokees in Transition: A Study of Changing Culture and Environment Prior to 1775.* 1977. ix + 207 p.
183. HAIGH, MARTIN J. *The Evolution of Slopes on Artificial Landforms, Blaenavon, U.K.* 1978. xiv + 293 p.
184. FINK, L. DEE. *Listening to the Learner: An Exploratory Study of Personal Meaning in College Geography Courses.* 1977. ix + 186 p.
185. HELGREN, DAVID M. *Rivers of Diamonds: An Alluvial History of the Lower Vaal Basin, South Africa.* 1979. xix + 389 p.

186. BUTZER, KARL W., ed. *Dimensions of Human Geography: Essays on Some Familiar and Neglected Themes.* 1978. vii + 190 p.
187. MITSUHASHI, SETSUKO. *Japanese Commodity Flows.* 1978. x + 172 p.
188. CARIS, SUSAN L. *Community Attitudes toward Pollution.* 1978. xii + 211 p.
189. REES, PHILIP M. *Residential Patterns in American Cities: 1960.* 1979. xvi + 405 p.
190. KANNE, EDWARD A. *Fresh Food for Nicosia.* 1979. x + 106 p.
192. KIRCHNER, JOHN A. *Sugar and Seasonal Labor Migration: The Case of Tucumán, Argentina.* 1980. xii + 174 p.
194. HARRIS, CHAUNCY D. *Annotated World List of Selected Current Geographical Serials, Fourth Edition. 1980.* 1980. iv + 165 p.
196. LEUNG, CHI-KEUNG, and NORTON S. GINSBURG, eds. *China: Urbanizations and National Development.* 1980. ix + 283 p.
197. DAICHES, SOL. *People in Distress: A Geographical Perspective on Psychological Wellbeing.* 1981. xiv + 199 p.
198. JOHNSON, JOSEPH T. *Location and Trade Theory: Industrial Location, Comparative Advantage, and the Geographic Pattern of Production in the United States.* 1981. xi + 107 p.
199-200. STEVENSON, ARTHUR J. *The New York–Newark Air Freight System.* 1982. xvi + 440 p.
201. LICATE, JACK A. *Creation of a Mexican Landscape: Territorial Organization and Settlement in the Eastern Puebla Basin, 1520-1605.* 1981. x + 143 p.
202. RUDZITIS, GUNDARS. *Residential Location Determinants of the Older Population.* 1982. x + 117 p.
203. LIANG, ERNEST P. *China: Railways and Agricultural Development, 1875-1935.* 1982. xi + 186 p.
204. DAHMANN, DONALD C. *Locals and Cosmopolitans: Patterns of Spatial Mobility during the Transition from Youth to Early Adulthood.* 1982. xiii + 146 p.
206. HARRIS, CHAUNCY D. *Bibliography of Geography. Part II: Regional. Volume 1. The United States of America.* 1984. viii + 178 p.
207-208. WHEATLEY, PAUL. *Nagara and Commandery: Origins of the Southeast Asian Urban Traditions.* 1983. xv + 472 p.
209. SAARINEN, THOMAS F.; DAVID SEAMON; and JAMES L. SELL, eds. *Environmental Perception and Behavior: An Inventory and Prospect.* 1984. x + 263 p.
210. WESCOAT, JAMES L., JR. *Integrated Water Development: Water Use and Conservation Practice in Western Colorado.* 1984. xi + 239 p.
211. DEMKO, GEORGE J., and ROLAND J. FUCHS, eds. *Geographical Studies on the Soviet Union: Essays in Honor of Chauncy D. Harris.* 1984. vii + 294 p.
212. HOLMES, ROLAND C. *Irrigation in Southern Peru: The Chili Basin.* 1986. ix + 199 p.
213. EDMONDS, RICHARD LOUIS. *Northern Frontiers of Qing China and Tokugawa Japan: A Comparative Study of Frontier Policy.* 1985. xi + 209 p.
214. FREEMAN, DONALD B., and GLEN B. NORCLIFFE. *Rural Enterprise in Kenya: Development and Spatial Organization of the Nonfarm Sector.* 1985. xiv + 180 p.
215. COHEN, YEHOSHUA S., and AMNON SHINAR. *Neighborhoods and Friendship Networks: A Study of Three Residential Neighborhoods in Jerusalem.* 1985. ix + 137 p.
216. OBERMEYER, NANCY J. *Bureaucrats, Clients, and Geography: The Bailly Nuclear Power Plant Battle in Northern Indiana.* 1989. x + 135 p.

217-218. CONZEN, MICHAEL P., ed. *World Patterns of Modern Urban Change: Essays in Honor of Chauncy D. Harris*. 1986. x + 479 p.

219. KOMOGUCHI, YOSHIMI. *Agricultural Systems in the Tamil Nadu: A Case Study of Peruvalanallur Village*. 1986. xvi + 175 p.

220. GINSBURG, NORTON; JAMES OSBORN; and GRANT BLANK. *Geographic Perspectives on the Wealth of Nations*. 1986. ix + 133 p.

221. BAYLSON, JOSHUA C. *Territorial Allocation by Imperial Rivalry: The Human Legacy in the Near East*. 1987. xi + 138 p.

222. DORN, MARILYN APRIL. *The Administrative Partitioning of Costa Rica: Politics and Planners in the 1970s*. 1989. xi + 126 p.

223. ASTROTH, JOSEPH H., JR. *Understanding Peasant Agriculture: An Integrated Land-Use Model for the Punjab*. 1989. xiii + 173 p.

224. PLATT, RUTHERFORD H.; SHEILA G. PELCZARSKI; and BARBARA K. BURBANK, eds. *Cities on the Beach: Management Issues of Developed Coastal Barriers*. 1987. vii + 324 p.

225. LATZ, GIL. *Agricultural Development in Japan: The Land Improvement District in Concept and Practice*. 1989. viii + 135 p.

226. GRITZNER, JEFFREY A. *The West African Sahel: Human Agency and Environmental Change*. 1988. xii + 170 p.

227. MURPHY, ALEXANDER B. *The Regional Dynamics of Language Differentiation in Belgium: A Study in Cultural-Political Geography*. 1988. xiii + 249 p.

228-229. BISHOP, BARRY C. *Karnali under Stress: Livelihood Strategies and Seasonal Rhythms in a Changing Nepal Himalaya*. 1990. xviii + 460 p.

230. MUELLER-WILLE, CHRISTOPHER. *Natural Landscape Amenities and Suburban Growth: Metropolitan Chicago, 1970-1980*. 1990. xi + 153 p.